Willame de Araújo Luz

Scientometric Evaluation of Prostate Adenocarcinoma

Willame de Araújo Luz

Scientometric Evaluation of Prostate Adenocarcinoma

In the SCOPUS database from 2005 to 2015

ScienciaScripts

Imprint

Any brand names and product names mentioned in this book are subject to trademark, brand or patent protection and are trademarks or registered trademarks of their respective holders. The use of brand names, product names, common names, trade names, product descriptions etc. even without a particular marking in this work is in no way to be construed to mean that such names may be regarded as unrestricted in respect of trademark and brand protection legislation and could thus be used by anyone.

Cover image: www.ingimage.com

This book is a translation from the original published under ISBN 978-613-9-72939-5.

Publisher:
Sciencia Scripts
is a trademark of
Dodo Books Indian Ocean Ltd. and OmniScriptum S.R.L publishing group

120 High Road, East Finchley, London, N2 9ED, United Kingdom
Str. Armeneasca 28/1, office 1, Chisinau MD-2012, Republic of Moldova, Europe
Printed at: see last page
ISBN: 978-620-7-89498-7

To Our Lord and to my beloved family, I dedicate it.

ACKNOWLEDGEMENTS

As a person of faith, and I have a lot of it, I thank God, the almighty, who can do anything and make us good people, for always making me believe in myself, even in the face of the greatest difficulties and doubts about my own ability. Without him, I wouldn't have achieved everything I've experienced and overcome to date.

To Our Lady of Fatima, my enlightened saint, who used her light to guide me along the paths that life forces us to take, but with her help, I had the wisdom to face them.

*To my pillars, my parents, **Maria das Neves Araújo Luz** and Raimundo Nonato de Araújo Luz, who, with simplicity and dedication, without a trace of tiredness, made me realise that life is only worth living if we do good and that it is only through study that we can make a positive contribution to society.*

*To my siblings, **Valmir** (in memorian), **Denildo, Leila** and **Laise**, for their strength and support.*

*To my beloved nephews, who are like children to me, **Layla Eduarda, Ana Larissa** and **Thomas**, love that is beyond me.*

*To my friends, the best there are, especially Giledystone **Samuel,** for translating my summary.*

*To my advisor Prof Dr **Cláudio Carlos da Silva**, for his patience and understanding. Thank you for accepting to be my supervisor and for your teachings.*

To all the professors of the Master's in Genetics at PUC Goiás who have passed through me, I know the importance and dedication of each one.

*The secretary of the master's programme, **Alessandra Malta**, for her availability and guidance on our doubts.*

*To my friends from the 7ª class of the Master's programme in Genetics at PUC Goiás, it was great to meet you. Especially **Marina Machado**, for her friendship.*

"**A** *mind that is open to a new idea will never return to its original size.*"

Albert Eistein

2

SUMMARY

SUMMARY

Cancer is defined as a chronic multi-causal disease characterised by the uncontrolled growth of cells that invade tissues and organs and can spread to other parts of the body. It is a major public health problem worldwide, especially in developing countries, due to the rapid growth of the older population. This scientometric study focused on the evaluation of prostate adenocarcinoma using FISH as a cytomolecular tool for analysing the PTEN gene, in the SCOPUS database from 2005 to 2015. To do this, the study was surveyed using the keywords:*" roostatic neoplasms* and *PTEN* and *FISH* Different approaches were taken to analyse the scientific production of the articles included in this study, such as: type of publication (experimental or review), number of articles/year, authors, scientific areas, journals, impact factor of the journals that published the most, among others. As a result, it was observed that the number of publications over the period studied varied over the ten years. Journals belonging to developed countries show that they invest in cutting-edge research and are dominant in terms of their scientific prestige. Brazilian scientific production has grown significantly in recent years, but the intellectual, social and economic impact of the publications produced in Brazil is still low and has a lot of growing to do. This should be changed with greater financial investment from both the government and private institutions in the area of scientific production, thus improving the credibility of research institutions and our researchers. Therefore, it can be concluded that this scientometric analysis presented data from quantitative studies on prostate adenocarcinoma, highlighting the importance of each of the proposed objectives, seeking alternatives to improve the growth of science and the visibility of productions on this subject in the context of global scientific activity.

Keywords: prostate cancer, impact factor, scientometrics, scientific production

CHAPTER 1

INTRODUCTION

Cancer has been defined as a chronic multi-causal disease characterised by the uncontrolled growth of cells that invade tissues and organs and can spread (metastasise) to other regions of the body (WCRF, 1997). Dividing rapidly, these cells tend to be aggressive and uncontrollable, leading to the formation of tumours (accumulation of cells) and/or malignant neoplasms. On the other hand, a benign tumour means a localised mass of cells that multiply more slowly and resemble their original tissue, rarely posing a risk to patients' lives (BRASIL, 2015).

The presence of cancer in humanity has been known for millennia. However, records designating the cause of death as cancer only began to exist in Europe in the 18th century. Since then, there has been a steady increase in cancer mortality rates, which seemed to rise sharply after the 19th century with the arrival of industrialisation (WHO, 1998). Cancer has emerged as a major public health problem all over the world, especially in developing countries, due to the rapid growth of the older population (MACHADO et al., 2009).

The causes of cancer are varied and can be external or internal to the body, both of which are interrelated. External causes are related to the environment and the habits or customs of a social and cultural environment. Internal causes are most often genetically predetermined and are linked to the body's ability to defend itself against external aggressions. These causal factors can interact in various ways, increasing the likelihood of malignant transformations in normal cells. Of all cases, 80 to 90 per cent of cancers are associated with environmental factors, such as smoking, eating habits, alcoholism, sexual habits, medication, occupational factors and solar radiation. Environmental risk factors for cancer are called carcinogens. These factors act by altering the molecular structure of the cells' DNA (BRASIL, 2015).

Carcinogenesis is a complex process, still poorly understood, which occurs in multiple stages in which cells become malignant through a series of progressive and cumulative mutations. These mutations arise from lesions caused by the interaction of physical, chemical and/or biological agents with the genetic material of host cells. Although these cells have repair mechanisms that efficiently remove most of the lesions introduced into their DNA, a small proportion of them are not repaired or are repaired incorrectly. As a result, mutations arise. The process of neoplastic transformation begins when these mutations alter the function of genes that directly or indirectly regulate cell proliferation or survival, such as proto-oncogenes and tumour suppressor genes (MACLEOD, 2000).

According to the National Cancer Institute (INCA) (INCA, 2015), the process of carcinogenesis, i.e. the formation of cancer, generally takes place slowly and it can take several years for a cancerous cell to proliferate and give rise to a visible tumour. These altered cells then start to behave abnormally and multiply uncontrollably.

In the initiation stage, the cells are affected by carcinogens, which cause modifications to some of their genes. At this stage the cells are genetically altered, but it is not yet possible to detect a tumour clinically. In the promotion stage, the genetically altered cells, i.e. "initiated", suffer the effect of carcinogens classified as oncopromotors. The initiated cell is slowly and gradually transformed into a malignant cell. For this transformation to take place, long and continuous contact with the carcinogen promoter is necessary. Suspending contact with promoting agents often stops the process at this stage. The progression stage is characterised by the uncontrolled and irreversible multiplication of altered cells. At this stage, the cancer has already taken hold and progresses until the first clinical manifestations of the disease appear. The factors that promote the initiation or progression of carcinogenesis are called onco-accelerating agents or carcinogens, and these stages are shown in Figure 1 (ALMEIDA et al., 2005).

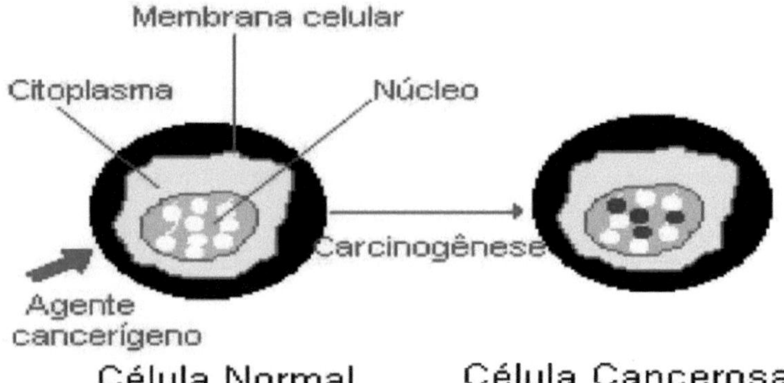

Figure 01. Schematic drawing of the process of transformation of a normal cell into a cancerous cell. Source: MINISTÉRIO DA SAÚDE (2015). NATIONAL CANCER INSTITUTE.

The prostate is a male reproductive gland that produces a fluid which, together with the spermatozoa, takes part in making semen. There are frequent pathologies in men's adult lives that can affect this gland, such as prostatitis, benign prostatic hyperplasia (BPH) and prostate adenocarcinoma (PCa) (BORGES DOS REIS & CASSINI, 2010).

Prostatitis is a commonly occurring condition that includes acute or chronic bacterial infection inside the prostate. It is theorised that chronic inflammation within the prostate due to exposure to microbial agents stimulates the production of inflammatory cytokines and reactive oxygen species, leading to increased cell proliferation and possibly carcinogenesis (CHENG, 2010).

Benign prostatic hyperplasia (BPH) is the most prevalent urological disease in men over the age of 50. Due to its high frequency and the expense of treating it, it is considered a public health problem in several industrialised countries. The two most well-known and well-studied determining factors in the development of BPH are age and androgens. BPH rarely occurs before the age of 30, and its incidence rises sharply after the age of 50 (FREIRE & PIOVESAN, 1999). From a histological point of view, BPH is characterised by hyperplasia of the stromal cells and epithelium of the prostate gland, resulting in an increase in its volume and the possibility of interference with the normal flow of urine caused by compression of the prostatic urethra and inadequate relaxation of the bladder neck (AVERBECK et al., 2010).

Adenocarcinoma of the prostate (CaP) is a malignant tumour of the prostate and is made up of different cell types, but in the overwhelming majority of cases it is a tumour called an adenocarcinoma. The cells that make up this tumour seem to have a similar origin to the cells that make up the normal prostate - glandular cells, but at a certain point they become more aggressive, multiplying more (or "dying" less) and increasing in number much faster than the rest of the cells in this organ. There is a classification of prostate tumours, based on their histological characteristics, which gives us an idea of how aggressive a particular tumour is. This classification is called the "*Gleason* Classification", *a* reference to the American doctor, a specialist in Pathological Anatomy, who studied thousands of cases of different individuals and drew up the classification currently used (IPIU, 2015).

In some European countries and the United States of America, epidemiological studies have shown an increase in the incidence of prostate cancer, particularly between the late 1980s and early 1990s, and a decline in mortality from the mid-1990s onwards (HALLAL et al., 2001; WUNSCH FILHO & MONCAU, 2002). On the other hand, in countries in Africa and Asia, incidence rates are still low. However, proportionally, more men die from the disease in these regions than in more developed countries (HAAS et al., 2008).

The International Agency for Research on Cancer (IARC) predicts that the number of deaths from prostate cancer in Brazil will almost double by 2015 (FERLAY et al., 2010). However, these projections are limited as there is no age or region data available. Trend analysis of cancer mortality are useful projections for planned interventions, such as screening programmes and advances in therapy, as well as for quantifying the likely incidence of cancer in the future (JEREZ- ROIG et al., 2014).

In the Centre-West region, mortality from prostate cancer has almost doubled in the last 30 years, making this the second Brazilian region with the highest mortality rate from this neoplasm. Geographical and socioeconomic factors, inherent to the different regions of the country, can make access to specialised cancer care services difficult and contribute to the fact that the time intervals for diagnosis and treatment show significant local variation (WUNSCH FILHO et al., 2008).

A closer look at the latest research into genetic services in Brazil (HOROVITZ, 2003) shows that diagnostic genetic tests were available in 47 (71%) of 66 genetic services linked to the Unified Health System (SUS). Among these, 83% offered conventional cytogenetics, 55% high-resolution cytogenetics, 32% fluorescent in situ hybridisation, 36% tests for inborn errors of metabolism and 32% prenatal diagnosis. Around 50% of the genetics services available also carry out research using molecular biology techniques for various groups of diseases, including mental retardation, dysmorphic syndromes, cancer predisposition, infertility, metabolic diseases, among others. There is no structured or published data on the number of genetic tests available in Brazil. A registry of genetic diseases is not available and most of the centres and care services related to the field of clinical genetics are concentrated in the Southeast and South, the most developed regions of the country. These services are generally integrated into university centres and referral hospitals and are responsible for the medical care of thousands of individuals and families every year. In addition, they are considered references at regional or national level (HOROVITZ et al., 2013).

In recent years, specialists and government authorities have become increasingly interested in quantitative indicators which, in addition to helping to understand the dynamics of science and technology, also serve as instruments for planning policies and making decisions in this sector. Scientometrics is a measuring device, based on statistical techniques, which aims to identify and process the information contained in scientific and technical publications available in information systems, essentially bibliographical references of articles, books and patents. Quantitative methods and especially data analysis are considered to be an indispensable element in advancing our understanding of the study of science as a means of producing and exchanging knowledge (SANTOS, 2003).

CHAPTER 2

LITERATURE REVIEW

2.1 Prostate

During the third month of embryonic development, the prostate develops from the epithelial invaginations of the posterior urogenital sinus. For this process to occur normally, the presence of 5a-dihydrotestosterone is necessary (HRICAK & SCARDINO, 2009). This molecule is synthesised from foetal testosterone by the action of 5a-reductase and is localised in the urogenital sinus and external genitalia of humans (WILSON et *al.*, 1981). Deficiencies of 5a-reductase will result in a rudimentary or undetectable prostate, as well as serious abnormalities of the external genitalia, although the epididymis, vas deferens and seminal vesicles remain normal (IMPERATO McGINLEY, 1975). During the prepubertal period, the constitution of the human prostate remains relatively identical. However, it undergoes morphological changes that reveal the adult phenotype with the onset of puberty. Ultimately, the gland increases in size to reach an average adult weight of approximately 20 g at around 25-30 years of age (HRICAK & SCARDINO, 2009).

The prostate is the largest accessory gland in the male reproductive system. It secretes a thin, slightly alkaline fluid that forms part of the seminal fluid. The prostate is made up of glandular elements and a stroma that are firmly fused within the prostatic capsule. The nerve supply to the prostate is derived from the prostatic plexus and the arterial supply from branches of the internal iliac artery. Lymphatic drainage from the prostate occurs predominantly via the internal iliac nodes (BHAVSAR & VERMA, 2014). The inner layer of the prostatic capsule is composed of smooth muscle with an outer collagenous lining (LEE et al., 2011).

The prostate is divided into four regions, the central zone (CZ), transition zones (TZ), peripheral zone (PZ), and anterior fibromuscular stroma (AFS) (BHAVSAR & VERMA, 2014). It consists of a base, apex, anterior face and two infero-lateral faces, as shown in Figure 2. In the upper part, its base is continuous with the bladder neck; in the lower part, the apex of the prostate rests on the superior fascia of the urogenital diaphragm; and in the anterior part, its surface relates to the pubic symphysis, separated from it by extraperitoneal fat in the retropubic space. Posteriorly, it is closely related to the anterior surface of the rectum, separated from it by the rectoprostatic septum (PINTO & MACÉA, 2010). Classically described as "walnut-shaped", the prostate is conical in shape and surrounds the proximal urethra as it exits the bladder (LEE et al., 2011).

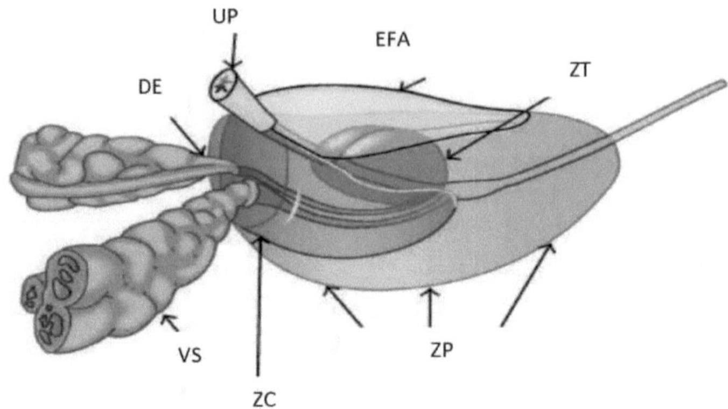

Figure 2. Zonal anatomy of the prostate gland. DE: ejaculatory ducts; UP: proximal urethra; VS: seminal vesicles; EFA: anterior fibromuscular stroma; ZT: transition zone; ZC: central zone; ZP: peripheral zone. SOURCE: Adapted from BHAVSAR & VERMA, 2014.

The PZ is the largest of the zones, comprising around 70 per cent of the glandular tissue. It extends from the base to the apex along the posterior surface and surrounds the urethra. In this zone, carcinoma, chronic prostatitis and post-inflammatory atrophy are relatively more common than in the other zones. The CCZ is located at the base of the prostate between the transitional and peripheral zones and accounts for approximately 25 per cent of the glandular tissue. It is a cone-shaped structure that surrounds the ejaculatory ducts and narrows to a vertex at the verumontanum, which is a longitudinal mucosal fold that forms an elliptical segment of the prostatic urethra, marking the point at which the ejaculate ducts enter the urethra. The ZT makes up only 5% of the glandular tissue and consists of two small lobes of glandular tissue that surround the proximal prostatic urethra just superior to the verumontanum. This is the portion of the glandular tissue that expands due to benign prostatic hyperplasia. The anterior EFA constitutes the convexity of the anterior external surface and is devoid of glandular tissue and instead composed of fibrous and smooth muscle elements. The apical half of this area is rich in striated muscle that mixes with the gland and the pelvic diaphragm muscle. As it extends laterally and posteriorly, it thins to form the fibrous capsule that surrounds the prostate gland (BHAVSAR & VERMA, 2014; PINTO & MACÉA, 2010). Although the term "capsule" is incorporated into current literature, there is no consensus on the presence of a true capsule (AYALA et al., 1989).

The zones have different embryonic origins and can be distinguished by their appearance, anatomical landmarks, biological functions, and susceptibility to pathologies; this information, including the composition of each region, is summarised in Table 1. Approximately 70 per cent of all prostate cancers arise from the PVZ, which is mainly derived from the urogenital sinus. In contrast, a higher incidence of PCa is found in the ZC. ZT shares the same embryological origin as ZP; however, the percentage of prostate cancers originating from ZT is lower. This can be explained by the differences in the stromal component of these two zones. The stroma of the TZ is more fibromuscular, and it has been postulated that BPH, which arises predominantly in the TZ, is a fibromuscular stromal disease (BHAVSAR & VERMA, 2014).

Table 1. Table showing the histological composition and embryonic origins of the various zones of the prostate.

	Central Zone (ZC)	Transition Zone (TZ)	Peripheral Zone (ZP)
Normal Prostate Volume (%)	25	5	70
Embryological Origin	Wolff duct	Urogential sinus	Urogenital sinus
Epithelium	Complex, large polygonal glands	Simple, small rounded glands	Simple, small rounded glands
Origin of CaP (%)	5	25	70
HPB (%)	-	100	-

Source: Adapted from BHAVSAR & VERMA, 2014.

The prostate is made up of three fundamental types of tissue: the epithelial, which forms the prostatic acini; the smooth muscle, whose fibres penetrate the interior of the gland and participate in the acinar emptying mechanism; and the connective stroma, which supports the entire structure of the organ (FREIRE & PIOVESAN, 1999).

Histologically, as shown in Figure 3, the adult prostate is made up of several prostatic alveoli, irregularly shaped secretory units that open through separate branched ducts into the prostatic urethra. The alveoli are embedded in a fibromuscular stroma with large amounts of collagen fibres mixed with bundles of irregularly distributed smooth muscle fibres. The stroma is continuous with the capsule and forms distinct lobes. The prostate contributes around 15 per cent of the ejaculate fluid. The secretory nature of the glandular epithelium of the alveoli is evident at high magnification: the pseudostratified epithelium has basal cells and secretory cells. The secretory cells, cylindrical or cuboidal in shape, produce a whitish serous liquid containing acid phosphatase, citric acid, zinc, prostate-specific antigen (PSA), and other proteases such as fibrinolytic enzymes involved in semen liquefaction. PSA is a serine protease that is quantified in the diagnosis of prostate diseases (NETTER et al., 2014).

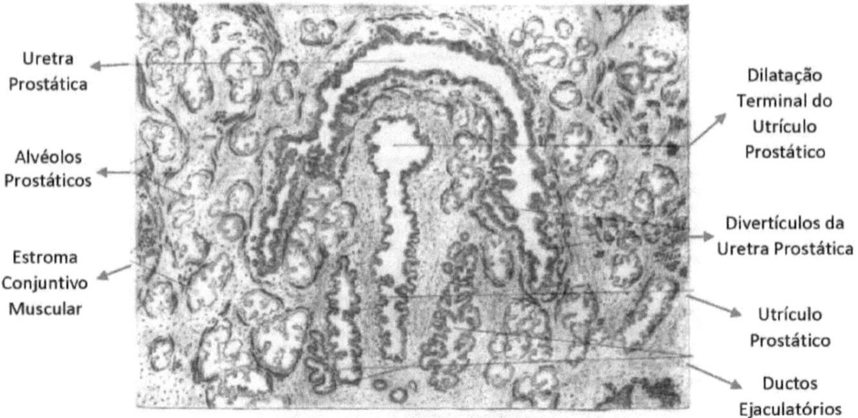

Figure 3 - Transverse section of the prostate: verumontanum, stained in haematoxylin-eosin (HE) at 80X magnification. Source: Adapted from ATLAS DE HISTOLOGIA DI FIORE (2000).

2.2 Prostate adenocarcinoma (CaP)

PCa is the most common solid tumour in men and one of the leading in terms of mortality (ZEQUI & CAMPOS; 2010). In recent years, PCa has become one of the most important and controversial topics in contemporary urology. Early detection campaigns and the dissemination of innovative knowledge have increased interest in this visceral neoplasm, recognised as the most frequent among men over 50. In this sense, technological innovations can

bring benefits in treatment, diagnosis and risk estimation when correctly incorporated into clinical practice and public health programmes. The expected benefit is a reduction in mortality from PCa. Possible harms include false-positive results, infections and bleeding resulting from biopsies, anxiety associated with the diagnosis of cancer and damage resulting from the treatment of tumours that would never evolve clinically (INCA, 2013; POMPEO, 1999).

Diagnosis in the early stages has been established in increasing incidence, a phase in which the chances of cure, or at least control, are much greater. Although PCa generally has a slow evolution, its natural history can be very variable, sometimes presenting early onset of metastases, at which stage cure becomes exceptional. It follows, therefore, that treatment must be instituted quickly (POMPEO, 1999).

Despite its high prevalence, the molecular mechanisms underlying the initiation and progression of prostate cancer are largely unknown, due to the great heterogeneity of the tumour (McCALL et al., 2008; HAN et al., 2009). The identification of molecular markers not only allows reliable prediction of the pathological stage of the disease, but prognosis is recognised as critically important in the future clinical management of prostate cancer and can be useful in the process of selecting therapeutic targets. Despite the clinical utility of the Gleason score, pathological stage and serum PSA to assess prognosis and guide causal management, but more determinant are the molecular markers that are needed to more precisely address the pathways that underlie PCa tumourigenesis (YOSHIMOTO et al., 2008; CHOUCAIR et al., 2012; FELGUEIRAS et al., 2014).

Molecular markers are useful tools in supporting various patient management measures: prevention, screening, diagnosis, prognosis, predicting treatment efficacy and monitoring treatment responses. Establishing a specific panel of markers, in any tissue or body fluids, can complement routine applying molecular diagnostic techniques to achieve earlier and more accurate diagnoses (FELGUEIRAS et al., 2014). This is a major challenge, as PCa is a silent disease in the early stages and therefore has no symptoms until it becomes locally advanced or metastatic (SMITH et al., 2003).

The current challenge facing prostate cancer researchers is to discover the critical genes and molecular pathways responsible for the onset of the neoplasm and the progression of the disease, as well as to develop new therapeutic strategies based on these discoveries. Initial progress in understanding the genetics of PCa began with cytogenetic and genomic analyses of primary tumours. Chromosomal losses involving 10q23.3 suggested that the PTEN gene (Phosphatase and Tensin Homolog Deleted on Chromosome Ten) may be a tumour suppressor, being one of the most common somatic genetic aberrations involved in the development of PCa and is often associated with high-risk disease (QI et al., 2014; LOTAN et al., 2011; YOSHIMOTO et al., 2007). PTEN deletion occurs in 20-70% of PCa and has been associated with rapid tumour progression and early recurrence (QI et al., 2014).

Most studies on PTEN loss in prostate cancer have focused on genomic deletions involving 10q.23.3. Such deletions, most frequently identified by *fluorescense* in *situ* hybridisation (FISH), occur in 10-70% of prostate cancer cases, depending on the study population examined, and are associated with a poor prognosis. For clinical situations, the assessment of *PTEN* allelic loss by FISH is somewhat complex, requiring the counting of the number of fluorescent signals in relation to control signals in interphase cells obtained by biopsy (LOTAN *et al.*, 2011).

PCa can take on different histological patterns, but most cases correspond to acinar adenocarcinomas that arise from prostatic epithelial cells that express androgen receptors (BOSTWICK, 1989). On the other hand, mucinous ductal adenocarcinomas, carcinomas and signet ring adenocarcinomas are extremely rare (GRIGNON, 2004).

The severe abnormalities of differentiation and proliferation that underpin the development of PCa can involve multiple genetic alterations, such as loss of heterozygosity, activation of oncogenes and loss of tumour suppressor genes (FOSTER *et al.*, 2000). But most are a consequence of natural ageing, where the process of aging is the most significant risk factor for the development of PCa (FELGUEIRAS *et al.*, 2014).

PCa is prone to metastasis as a result of various molecular mechanisms. In general, these processes lead to local invasion, migration and site-specific establishment of metastases in secondary locations, usually in the bones, lungs or liver (BUBENDORF *et al.*, 2000).

2.3 Diagnosis of Prostate Adenocarcinoma

Screening for prostate cancer is carried out using rectal examination, the Prostatic Acid Phosphatase (PAP) test and the Prostate Specific Antigen (PSA) test, which includes preoperative PSA levels. However, they have limitations in terms of sensitivity and specificity and low positive predictive value (THOMPSON et al., 2005; BRYANT & HAMDY, 2008). As a result, the benefits and risks of screening for this type of tumour have been widely debated in the literature and there is no consensus on guidelines for its use at a population level (BRYANT & HAMDY, 2008; CRAWFORD & ABRAHAMSSON, 2009; GREENE et al., 2009; SCHODER et al., 2009). A model for predicting disease progression includes preoperative PSA levels and the Gleason score of the surgical specimen, when the disease is organ-confined (PARTIN et al., 1995).

One of the ways pointed out in the literature for screening is the rectal examination, a low-cost and quick procedure (SOUZA et al., 2011). It is used to assess the size, shape and consistency of the prostate in order to check for the presence of nodules, but it is known that this examination has some limitations, since it only allows palpation of the posterior and lateral portions of the prostate, leaving 40% to 50% of tumours out of reach; it depends on the training and experience of the examiner and there is still resistance and rejection from a significant proportion of patients in relation to this type of examination (NAGLER et al., 2005). Despite the facilities, it stirs up the male imagination and is interpreted as an affront to masculinity, which can influence adherence to the test (GOMES et al., 2008).

PAP is a glycoprotein dimer produced predominantly by the prostate and was initially used in serum as a marker for the detection of metastatic PCa (GUTMAN & GUTMAN, 1938). Unfortunately, PAP has low sensitivity for detecting localised disease (HERNANDEZ & THOMPSON, 2004) and was replaced as a test following the discovery and development of the PSA antigen test (HARA et al., 1971).

PSA is a 33 kDa serine protease (kallikrein-3) that is secreted by prostate epithelial cells. In the normal prostate, PSA is secreted from the prostate epithelium into the secretory ducts, contributing to the seminal fluid. However, in PCa, the rupture of the basal cell layer allows PSA to "leak" into the circulation, resulting in high levels of PSA in the blood (VELONAS et al., 2013). Its high level in the bloodstream is considered an important biological marker for some prostate diseases, including PCa (CARROL et al., 2009; THOMPSON & ANKERST, 2005; WOLF et al., 2010).

The mechanism of hormonal regulation of kallikreins has been studied in depth. The PSA regulatory gene is related to androgens. Therefore, drugs that affect the production or metabolism of androgens influence serum PSA levels. Finasteride (at doses of 1 mg or 5 mg a day) reduces the PSA value by 50 per cent six months after starting treatment, while Dutasteride takes twelve months to achieve this reduction. The concept that tumour cells produce more PSA is not correct; the pathophysiology of the increase in the plasma concentration of PSA is based on the occurrence of cell lysis, enabling its release into the bloodstream. PSA is an organ-specific marker and not disease-specific. Three of the most common prostate conditions can raise it: prostatitis, BPH and PCa. Treatment with antibiotics can reduce the level of elevated PSA secondary to prostatitis by approximately 30 per cent (BORGES DOS REIS & CASSINI, 2010).

PSA does not distinguish between the stages of PCa and, significantly, does not identify metastatic PCa with the sensitivity and specificity required to make accurate therapeutic decisions (HESSELS & SCHALKEN, 2013). The widespread use of PSA as a screening tool has been partly responsible for the rapid increase in PCa diagnoses over the last two decades. Although mortality associated with PCa has decreased over the years, it is uncertain whether this is due to the introduction of the PSA test or to the advances and effectiveness of current PCa treatments (SARDANA et al., 2008).

In order to improve the sensitivity (percentage of men with the disease in whom there is a change in PSA) and specificity (percentage of men without the disease in whom PSA remains unchanged) of serum PSA dosage for diagnosing prostate cancer, new parameters have been introduced using PSA isoforms, serum PSA level, prostate volume, adjustment for age and elevation kinetics (BORGES DOS REIS & CASSINI, 2010).

Total PSA dosage cannot be used in isolation as a predictive factor of tumour extension in the prostate gland or the presence of metastases, but it does provide important information that can be used when deciding on the therapy to be employed. Approximately 80% of prostate tumours are confined to the gland when PSA values are below 4.0 ng/mL. When the PSA is between 4.0 and 10.0 ng/mL, 66% of patients have confined tumours, but when it is above 10.0 ng/mL, the chance of tumours without extraprostatic extravasation is approximately 35%. Lymph node metastases occur in around 20% of patients with a PSA > 20 ng/mL and in 75% of patients with a PSA > 50 ng/mL. The higher the PSA value, the greater the chance of locally advanced or disseminated disease. This has a major impact on the therapeutic decision and prognosis of the disease (BORGES DOS REIS & CASSINI, 2010). Table 2 shows the PSA level in relation to the patient's age.

Table 2. Mean and maximum PSA values in relation to age.

Age (in years)	Mean PSA value (ng/mL)	Maximum value PSA (ng/mL)
40-50	0,7	2,5
50-60	1,0	3,5
60-70	1,4	4,5
>70	2,0	6,5

Source: BORGES DOS REIS, R. & CASSINI, M.F.2010.

Unfortunately, the PSA test can result in false positives (VELONAS et al., 2013). It also falsely identifies indolent PCa, causing 40% - 50% of cases to be treated unnecessarily (LIN et al., 2013). Reducing the PSA threshold has been suggested as an alternative to satisfy the current problems of the PSA test. However, when the PSA threshold is lowered there is an increased risk of identifying and treating indolent disease unnecessarily. Furthermore, a PSA > 4 ng/mL can usually be caused by benign prostatic hyperplasia, prostatitis and rarely by other human malignancies (BODEY et al., 1997; HAYTHORN & ABLIN, 2011).

After radical prostatectomy, it takes an average of three months for the PSA to reach undetectable or very low levels (< 0.04 ng/mL). Post-surgical biochemical recurrence is currently defined when the PSA value, after reaching undetectable levels, rises again and exceeds 0.20 ng/mL. This is the only situation in which total PSA is 100% sensitive and specific. Early biochemical recurrence (< 6 months) suggests advanced (metastatic) disease, while a late increase in PSA (> 1 year) suggests local recurrence. When the serum PSA level reaches undetectable levels after surgery, it should be interpreted as the presence of residual local or metastatic prostate tissue. Often, imaging methods cannot identify the presence of this local residual tissue or micrometastases (BORGES DOS REIS & CASSINI, 2010).

The Gleason score is a powerful prognostic marker and correlates with the extent of the disease, particularly with the risk of extraprostatic involvement, making it the strongest clinical predictor of PCa progression (GLEASON et al., 1979; RUBIN et al., 2000; CORRÊA et al., 2006). The classification system, first described in 1960, characterises prostate tumour architecture and morphology. The pathologist assigns a primary and secondary score, and the total score is totalled over a range of 2 to 10 (MARTIN et al., 2011). Tumours with a Gleason score of 7 or more are considered to be biologically aggressive, tumours with a score of 5 or 6 are considered to be intermediately aggressive, and tumours with a score between 2 and 4 are considered to be biologically less aggressive (KOCH et al., 2000).

2.4 Clinical Staging of Prostate Adenocarcinoma

Despite the increase in epidemiological and molecular knowledge about prostate cancer, it is impossible to predict which patients will develop clinically significant disease and which will remain with a confined tumour. Therefore, early detection of PCa has allowed many patients to have the possibility of radical surgical treatment with curative intent. In this way, histopathological analysis has great clinical relevance, as well as histological and serological examinations describing a large number of important alterations, allowing the evolution of the disease to be monitored (NASSIF & FILHO, 2010).

The TNM system for classifying malignant tumours was developed by Frenchman Pierre Denoix between 1943 and 1952. In 1950, the International Union Against Cancer (UICC) appointed a Committee on Nomenclature and Statistics of Tumours and adopted, as the basis for its work on the classification of clinical staging, the general definitions of local extension of malignant tumours suggested by the Sub-Committee on Cancer Case Registries and Statistical Presentation of the World Health Organisation (WHO) (BRASIL, 2004).

Knowledge of the biology of tumours led the UICC to develop a system for classifying the evolution of malignant neoplasms in order to determine the best treatment and patient survival. This system, known in Brazil as "staging", is based on assessing the size of the primary tumour (represented by the letter T), the extent of its spread to regional lymph nodes (represented by the letter N) and the presence or absence of distant metastasis (represented by the letter M) and is known as the TNM System for the Classification of Malignant

Tumours (BRASIL, 2015).Each clinical staging category has several subcategories: for the primitive tumour, they range from T1 to T4; for lymphatic involvement, from N0 to N3; and for metastases, from M0 to M1 - with some tumours not necessarily fulfilling all the T or N categories. The combination of the various TNM subcategories (letters and numbers) determines the clinical stages, which range from I to IV in most cases. Clinical staging therefore also represents the language available to the oncologist to define behaviour and exchange knowledge based on data from the physical examination and complementary tests relevant to the case (BRASIL, 2015). Table 5 shows the staging of PCa.

Table 3. International consensus proposal for the staging of PCa.

TNM	DEFINITION	TNM	DEFINITION
T0	No evidence of prostate tumour	T_{3b}	Bladder neck invasion
T1a	Non-palpable tumour < 5%	T_{C3}	Invasion of the seminal vesicles
T1b	Non-palpable tumour > 5%	T_4	Invasion of the pelvic wall
T1c	Non-palpable tumour, Altered PSA	No	No lymph node metastases
T2a	Smaller nodule at $< /^1{}_2$ lobe	Ni	Metastases in iliac lymph nodes
T2b	Nodule > *Vi* lobe	N_2 and	
		N_3	Metastases in aortic lymph nodes
T2c	Bilateral nodule	Mo	No systemic metastases
T3a	Minimum perprostatic extension	M1	Systemic metastases

Source: BORGES DOS REIS, R. & CASSINI, M.F.,2010.

Pathological staging is based on the surgical findings and pathological examination of the surgical specimen. It is established after surgical treatment and determines the extent of the disease with greater precision. This staging may or may not coincide with the clinical staging and is not applicable to all tumours. However, for some tumours (skin and ovarian, for example) it is the only possible staging. It is written with the lower case letter p before the letters T, N and M: Example: pT1pN1pM0 (BRASIL, 2015).

Given the limitations of clinical staging, new clinical prognostic factors have been intensively studied. Clinical prognostic factors are obtained prior to treatment and guide the choice of the best therapeutic option for each patient (JOHANSSON et al., 2004). The most important clinical prognostic markers currently available are pre-treatment PSA levels and the degree of histological differentiation of the biopsied tumour fragments, according to the Gleason classification. Extensive literature supports the clinical usefulness of these prognostic factors (BOSTWICK et al., 2000).

Many tumours clinically classified as localised are in fact not, leading to ineffective curative therapeutic indications (HAESE et al., 2000). On the other hand, patients with tumours of low clinical significance are treated unnecessarily due to the current limitations of the prognostic classification (ANDRÉN et al., 2006).

Inaccuracy in defining the pre-treatment prognosis of localised prostate cancer is a serious public health problem, given the high morbidity associated with the treatment options commonly used (MIGOWSKI & AZEVEDO E SILVA, 2010). For these authors, biochemical recurrence and pathological staging (after surgical treatment) are the most commonly used outcomes in studies on clinical prognostic factors in prostate cancer. Important outcomes such as the development of metastases and specific mortality are not widely used due to the need for long follow-up times.

2.5 Epidemiology of PCa

PCa is a major public health problem and an important health problem for the male population. PCa is the most commonly diagnosed cancer in men (LOTAN et al., 2011). According to the WHO, the distribution of cancer deaths around the world is not homogeneous. Epidemiologists who study cancer have observed that its prevalence in the world has increased significantly over the last century. It is believed that this result is related, among other things, to the industrialisation and urbanisation that took place during this period. In fact, the morbidity and mortality associated with cancer observed in developed countries is higher than in developing countries. Furthermore, some specific forms of cancer, such as colon and rectal cancer, prostate cancer and female breast cancer, are more frequent in developed countries, while others, such as stomach, oesophageal and cervical cancer, have a higher incidence in developing countries. Studies show that different cancer patterns are also observed among individuals who emigrate to a new country or region (GARÓFOLO et al., 2004).

PCa is the second leading tumour in terms of incidence and the sixth leading cause of death among men worldwide (FERLEY et al., 2010). The highest rates of PCa are found in Europe, North America and Australia. In recent years, recorded incidence rates are increasing in many regions, partly related to advances in diagnosis. Its incidence increases particularly after the age of 60-70 (PETO, 2001).

According to global estimates from the Globocan 2012 project of the WHO's International Agency for Research on Cancer (IARC), there were 14.1 million new cases of cancer and a total of 8.2 million cancer deaths worldwide in 2012. It is estimated that by 2030, the global burden will be 21.4 million new cancer cases and 13.2 million cancer deaths, as a result of population growth and ageing, as well as a reduction in infant mortality and deaths from infectious diseases in developing countries (BRASIL, 2014).

In Brazil, the estimate for 2014, which will also be valid for 2015, points to the occurrence of approximately 576,000 new cases of cancer, including non-melanoma skin cancers, reinforcing the magnitude of the cancer problem in the country. Non-melanoma skin cancer (182,000 new cases) will be the most common cancer in the Brazilian population, followed by prostate cancer (69,000), female breast cancer (57,000), colon and rectum cancer (33,000), lung cancer (27,000), stomach cancer (20,000) and cervical cancer (15,000). Without considering cases of non-melanoma skin cancer, 395,000 new cases of cancer are estimated, 204,000 for males and 190,000 for females (BRASIL, 2014).The proportional distribution of the ten most incident types of cancer estimated for 2014 by sex in Brazil, except non-melanoma skin, is shown in Figure 4.

Localização primária	casos	%	Homens	Mulheres	Localização primária	casos	%
Próstata	68.800	22,8%			Mama Feminina	57.120	20,8%
Traqueia, Brônquio e Pulmão	16.400	5,4%			Cólon e Reto	17.530	6,4%
Cólon e Reto	15.070	5,0%			Colo do Útero	15.590	5,7%
Estômago	12.870	4,3%			Traqueia, Brônquio e Pulmão	10.930	4,0%
Cavidade Oral	11.280	3,7%			Glândula Tireoide	8.050	2,9%
Esôfago	8.010	2,6%			Estômago	7.520	2,7%
Laringe	6.870	2,3%			Corpo do Útero	5.900	2,2%
Bexiga	6.750	2,2%			Ovário	5.680	2,1%
Leucemias	5.050	1,7%			Linfoma não Hodgkin	4.850	1,8%
Sistema Nervoso Central	4.960	1,6%			Leucemias	4.320	1,6%

*Números arredondados para 10 ou múltiplos de 10.

Figura 4.   Proportional distribution of the most incident types of cancer estimated for 2014, except non-melanoma skin.
Source: Estimate 2014: Cancer Incidence in Brazil/National Cancer Institute José Alencar Gomes da Silva.

In the Centre-West Region, in the state of Goiás, in its capital Goiânia, the estimated gross incidence rates for 2014, according to state and capital, estimated per 100,000 inhabitants and the number of new cancer cases, according to sex and primary location are shown in Figure 5.

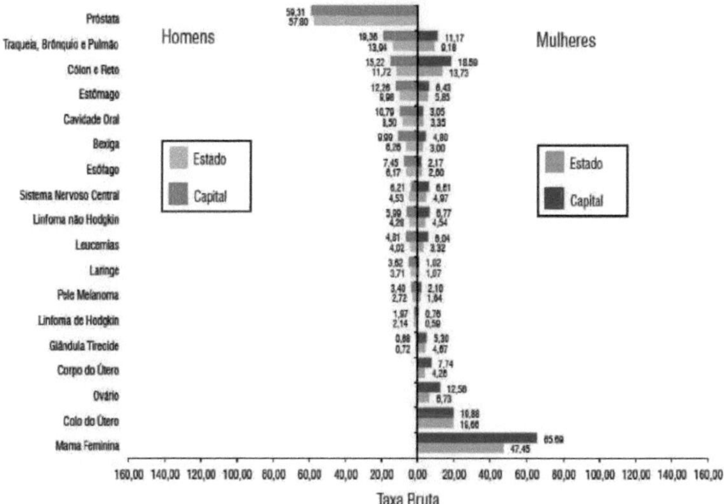

*Valores por 100 mil habitantes.

Figura 5.   Estimated gross incidence rates for 2014 by sex, by state and capital, Goiás.
Source: Estimate 2014: Cancer Incidence in Brazil/National Cancer Institute José Alencar Gomes da Silva.

An estimated 68,800 new cases of prostate cancer are expected in Brazil in 2014, as shown in Figure 6. These figures correspond to an estimated risk of 70.42 new cases per 100,000 men. Without considering non-melanoma skin tumours, prostate cancer is the most common cancer among men in all regions of the country, with 91.24/100,000 in the South, 88.06/100,000 in the Southeast, 62.55/100,000 in the Midwest, 47.46/100,000 in the Northeast and 30.16/100,000 in the North (BRASIL, 2014).

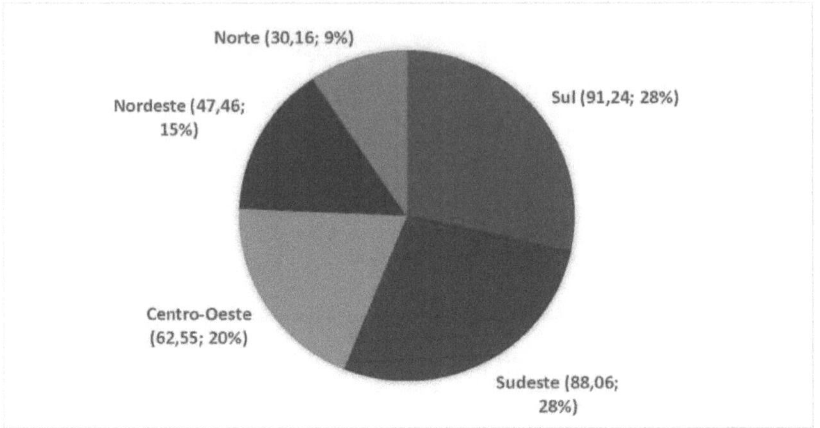

Figure 6: Incidence and percentage of prostate cancer in Brazilian regions. SOURCE: Adapted from INCA 2014.

The only well-established risk factor for developing PCa is age. Approximately 62 per cent of cases diagnosed worldwide occur in men aged 65 or over. With the increase in life expectancy worldwide, the number of new cases of prostate cancer is expected to rise by around 60 per cent by the year 2015. In addition, ethnicity and family history of the disease are also considered risk factors. PCa is approximately twice as common in black men compared to white men. American, Jamaican and Caribbean men of African descent have the highest incidence rates of prostate cancer in the world, which can be attributed in part to heredity (around 5% to 10%). However, it is possible that this difference between blacks and whites is due to lifestyle or factors associated with the detection of the disease (BRASIL, 2014).

Another important factor in the aetiology of this type of cancer is diet. Diets based on animal fat, red meat, sausages and calcium have been associated with an increased risk of developing prostate cancer. Obesity has also been shown to increase the risk of developing this neoplasm, especially those with more aggressive behaviour. On the other hand, it is possible that diets rich in vegetables, vitamins D and E, lycopene and omega-3 may be able to confer some protective effect against prostate cancer (BRASIL, 2014).

2.6 Etiology and risk factors of PCa

The etiology of PCa is unknown, although some risk factors have been identified: genetic, racial and dietary factors (RHODEN & AVERBECK, 2010).

Epidemiological evidence suggests that prostate cancer has a relevant genetic and familial component. In this context and from a phenotypic point of view, this neoplasm is classified as: (i) sporadic prostate adenocarcinoma, (ii) familial and (iii) hereditary. Sporadic cancers (85%) occur in individuals with a negative family history. Familial PCa is defined as the occurrence of this condition in a man with 1 or more family members affected by the disease. A small population of individuals (around 9%) has true hereditary prostate cancer, defined by three or more affected family members, the occurrence of this condition in 3 successive generations or at least two family members diagnosed with the disease before the age of 55 (CARTER et al., 1992).

If a first-degree relative has the disease, the risk of cancer in the men of that family is at least twice as high when compared to other men in the general population. If two or more first-degree relatives are affected, the risk increases 5 to 11 times (AUS et al., 2001; GRONBERG et al., 1996), as shown in Table 4.

Table 4: Family history in relation to relative and absolute risk of PCa.

FAMILY HISTORY	RELATIVE RISK	ABSOLUTE RISK
NONE	1	8
FATHER OR BROTHER	2	15
FATHER AFFECTED < 60 YEARS	3	20
HEREDITARY CANCER	5	35-45

Source: Adapted from AUS, et al., 2001.

Hereditary factors are important in the development of clinical PCa and exogenous factors can have an important impact on increasing the risk. The question is whether or not there is sufficient evidence to recommend lifestyle changes to reduce the risk (SCHULMAN et al., 2000).

Age is an important risk factor for prostate cancer. Prostate cancer is rarely seen in men under 40, the incidence increasing rapidly with each subsequent decade. For example, the probability of being diagnosed with prostate cancer is 1 in 304 for men aged < 49; 1 in 44 for men aged 50 to 59, 1 in 16 for men aged 60 to 69, and 1 in 9 for men aged > 70 (A.C.G., 2015). The risk of developing and dying from prostate cancer is significantly higher among blacks, and at intermediate levels among whites, and is lowest among native Japanese (ALTEKRUZE et al., 2010; BUNKER et al., 2010).

Genetic instability is also an important risk factor. Every man is born programmed to have prostate cancer, as they all carry proto-oncogenes in their genome, which signal a normal cell to turn malignant. Prostate cancer arises because multiple cell divisions occur over the years, accompanied by discrete losses of suppressor genes and activation of proto-oncogenes, due to inflammation or the influence of local mediators (SROUGI et al., 2008; CARVALHO et al., 2009; RHODEN & AVERBECK, 2010).

Endogenous hormones, including androgens and estrogens, probably influence prostate carcinogenesis. It has been widely reported that eunuchs and other individuals with castrate levels of testosterone before puberty do not develop prostate cancer (WU & GU, 1991).

Some dietary risk factors may be important modulators of prostate cancer risk. These include consumption of fat and/or red meat, lycopene (KOLONEL, 2001) and dairy products/calcium/vitamin D. Phytochemicals are non-nutritive compounds derived from plants, and it has been proposed that phytoestrogens in the diet may play a role in preventing prostate cancer (CHAN & GIOVANNUCCI, 2001).

Obesity related to PCa affects substantial proportions of the male population, and the association between the two is of great public health importance (ALLOTT et al., 2013). Therefore, evidence suggests that obesity (either before or at the time of diagnosis) is strongly associated with prostate cancer-specific progression and mortality, independent of lifestyle or clinical factors (CHAN et al., 2014).

Smoking is associated with a moderate increase in the risk of prostate cancer (HUNCHAREK et al., 2010). And this association is very evident, since in smokers there is an increase in aggressive and fatal cancers, suggesting that smoking may be involved in metastatic promotion (ZU & GIOVANNUCCI, 2009).

2.7   PTEN gene as a molecular marker in CaP

Among the genes studied in CaP, PTEN stands out, which is a tumour suppressor gene (GST), located at 10q23.3 as shown in Figure 7, with the function of encoding a phosphatase that participates in the regulation of the cell cycle in the G1 phase, cell adhesion and apoptosis, and can be inactivated as a result of mutations and deletions in various solid neoplasms, including CaP (RISINGER et al, 1997; VISAPAA et al., 2003; CAMPOS et al., 2013). According to Leslie & Downes (2004), PTEN is a 403 amino acid protein and a member of the large protein tyrosine phosphatase (PTP) family.

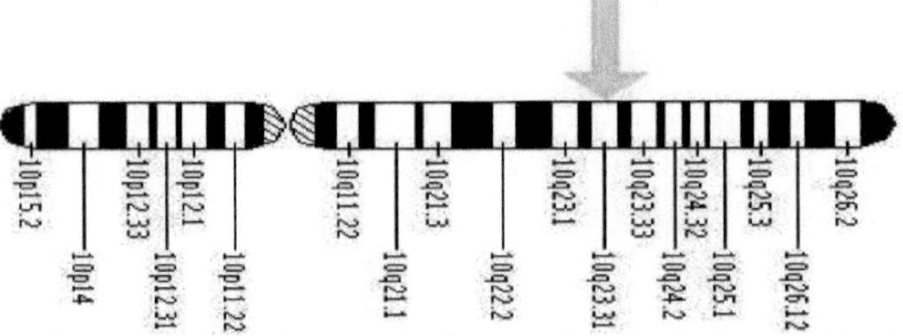

Figure 7. Ideogram of the human chromosome banding pattern, indicating the chromosomal location of the PTEN gene at 10q23.3.
Adapted from: http://ghr.nlm.nih.gov/dynamicImages/chromomap/PTEN.jpeg. (Medicine 2007).

PTEN is one of the most commonly lost tumour suppressor genes in human cancers, and its deregulation is also implicated in several other diseases (LESLIE & DOWNES, 2004). Several mutations have been associated with prostate cancer (GRAY et al., 1998), PTEN protein and genomic losses have been correlated with a poor prognosis in PCa (McMENAMIN et al., 1999; SCHMITZ et al., 2007; YOSHIMOTO et al., 2007).

Another important role of PTEN is the modulation of phosphatidylinositol-3-kinase (PI3K), a pathway downstream of the protein kinase (AKT), the Murine Thymonna Viral Oncogene Homolog. This pathway regulates a series of target genes that are involved in apoptosis and cell cycle progression (BHALLA et al., 2013). Loss of PTEN and subsequent activation of the PI3K pathway are associated with tumour progression in prostate cancer (KOKSAL et al., 2004; BERTRAM et al., 2006). Deletion of PTEN is associated with an increase in transformation potential and marks an increase in cell proliferation in prostate cancer through its negative regulation of the PI3K pathway (HAN et al., 2009).

FISH studies allow robust assessment of genomic loss in tumour samples of reasonable size. Most studies have identified an association between PTEN genomic loss and worse clinical evolution of PCa (YOSHIMOTO et al., 2007; REID et al., 2012). Previous studies of PTEN by FISH have mainly used individual probes that cover the entire gene. The genomic probe of the PTEN gene spans 368 kb and starts 166 kb from the 5' end of the gene and extends 98 kb from the 3' end of the gene (QI et al., 2014).

Somatic alterations of PTEN include point mutations, small deletions, deletions and epigenetic silencing. The functional loss of PTEN and subsequent activation of the AKT pathway, one of the most common abnormalities in CaP progression, and chromosomal region mapping has shown PTEN to be the most frequently suppressed gene in CaP (MUGA et al., 2010). Different studies have indicated that PTEN haploinsufficiency could be an early prognostic marker for PCa, and complete loss of PTEN expression seems to correlate with advanced pathological stage and high Gleason score (LIU et al., 2006).

Although the ETS-related gene rearrangement and PTEN deletion are strongly implicated in the development of PCa, little is known about the link between these two genomic events. Yoshimoto et al. (2006) reported that the TMPRSS2-ERG fusion may be accompanied by PTEN deletion in localised prostate cancer. Using these markers as a combined panel would allow better differentiation between low risk and high risk prostate cancer (QU et al., 2013).

The molecular alterations involved in the pathogenesis and natural history of PCa are poorly understood. Thus, the essential steps that mark the transition from the initial stages of prostate cancer development to the more aggressive stages of the disease are not known. In the literature on PCa, mutations in PTEN have been described with discrepant data. Although mutations in PTEN were first reported in prostate cancer in 1997, relatively few studies have analysed the mutational status of this gene in prostate cancer since then, and perhaps the number of adenocarcinomas tested in these studies has been very small, representing an understudied event (MUGA et al., 2010).

It would be interesting to investigate other mechanisms of alteration in PTEN, such as deletion or epigenetic inactivation. The most common mechanism of PTEN inactivation of both alleles is mutation of one of them and deletion of the other, although reduced levels of PTEN proteins are often seen in the absence of genomic abnormalities (MUGA et al., 2010). Loss of heterozygosity (LOH) of PTEN is frequently reported in primary prostate adenocarcinomas, while the proportion of adenocarcinomas with deletion of one allele and mutation of the other is low (PARSONS, 2004; UZOH et al., 2009).

Several studies have indicated that haploinsufficiency of the PTEN protein could be an early prognostic marker for prostate cancer, and complete loss of expression seems to correlate with pathological markers of poor prognosis and tumour progression (MUGA et al., 2010).

The reported frequency of *PTEN* loss in prostate cancer varies widely, probably as a result of differences in tissue preparation, disease stage, and the methodology used to detect molecular aberrations. The heterogeneity of these studies potentially obscures the clinical impact of PTEN loss in human prostate cancer (YOSHIMOTO et al., 2006).

2.8   FISH - Fluorescent in situ hybridisation

Among the molecular cytogenetic techniques that are particularly useful for diagnoses in medical science, including gene mapping, diagnosis of chromosomal anomalies and studies of cell structure and function, the Fluorescense in situ Hybridisation (FISH) technique is generally the one of choice in most study and diagnostic centres.

In clinical research, FISH can be used for prenatal diagnosis of chromosomal aberrations and postnatal diagnosis of chromosomopathies (NATH et al., 2000). FISH is a powerful technique for locating specific DNA or RNA sequences in cells, tissues and tumours, in interphase chromatin and metaphase chromosomes to identify chromosomal, numerical and structural alterations. It provides a unique link between studies of cell biology, cytogenetics and molecular genetics (HOVHANNISYAN, 2010).The principle of the FISH technique methodology, as shown in Figure 8, is based on the ability of single-stranded DNA to hybridise with complementary DNA, i.e. a large number of copies of a particular nucleic acid segment (DNA or RNA) are directly marked with fluorochromes: fluorescein isothiocyanate (FITC), Texas red isothiocyanate (TRITC), rhodamine, spectrumOrange, spectrumGreen, among others or indirectly labelled with haptens (biotin, digoxigenin). Identified DNA segments are denatured, marked with specific fluorochromes and used as probes to recognise and hybridise with homologous sequences of denatured nuclear DNA fixed to microscope slides (HACKEL & VARELLA-GARCIA, 1997).

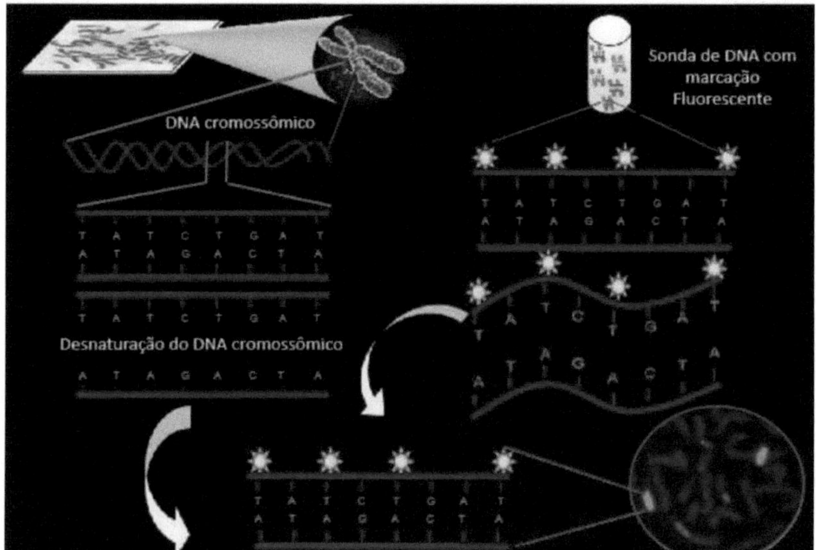

Figure 8. Simplified schematic representation of the FISH technique. Source: West Coast Pathology Laboratories, 2013.

The hybridised sequences are observed using fluorescence microscopy. The light from the ultraviolet (UV) source of a halogen lamp is filtered so that only the wavelengths suitable for the excitation of fluorescent molecules reach the sample. The light emitted by fluorochromes is generally of longer wavelengths, which makes it possible to distinguish between the excitation and emission of light by means of a second optical filter. It is therefore possible to see bright coloured signals on a dark background. It is also possible to distinguish between several excitation and emission bands, thus between several fluorochromes, which allows the observation of many different probes on the same cell target (NATH et al., 2000).

The FISH technique has a positive and relevant impact on genetic diagnosis and is a requirement for future molecular investigations, helping to identify genes related to the tumour phenotype. A major advantage of in situ hybridisation is that the test allows maximum use of tissue that is difficult to obtain (e.g. embryos and clinical biopsies). Several different hybridisations can be performed on the same tissue. Tissue libraries can be formed and stored in a freezer for future use (JENSEN, 2014).

The advantages of using FISH are related to the speed of research into aneuploidies, the fact that there is no need to grow cells in culture and that preserved tissues, formalin-fixed tissues, blood and bone marrow samples can be used. Of the various means of determining the genomic deletion of PTEN, FISH offers the advantage of being highly specific and quantitative, as well as allowing the number of copies of the gene to be determined within individual cells in tissue sections (YOSHIMOTO, 2006).

Although many technical problems still limit the current application of FISH to non-dividing cells, interphase cytogenetics with this methodology has become an increasingly appropriate approach for investigating important biological questions. Methodological advances in both sample and probe preparations, as well as in systems for acquiring and processing images and analysing hybridisation signals, contribute to the use of FISH in diagnostic procedures (HACKEL & VARELLA-GARCIA, 1997).

2.9 Scientometrics

With the development of scientific research and the constant dissemination of findings, it has become necessary to evaluate these advances and determine the developments achieved by the different areas of knowledge, thus giving rise to scientific meters, named according to their various activities, and within these scientific meters there is scientometrics (VANTI, 2002).

The term scientometrics originated in the former Soviet Union, becoming better known in the late 1970s with a publication in the journal *sCcieiil()melrK.\'* in Hungary. Academics became more interested in scientometrics in the 1980s due to the emergence of a database provided to universities by the former *"hislilule for Scientific Information"* (ISI, now Thomsom ISI) (VANTI, 2002). This database has information on the publications of various journals, in different approaches and in the most varied fields of knowledge (STREHL & SANTOS, 2002).

Scientometrics can be defined as the study of the quantitative aspects of science, by establishing it as a discipline or economic activity. This type of metric study of information is considered an area of the sociology of science, which covers quantitative analyses of scientific activities and identifies domains of interest where subjects are more concentrated (TAGUE-SUTCKIFFE, 1992; MACIAS-CHAPULA, 1998).

Scientometrics can be used to assess the importance of a particular subject, author and/or work, as well as highlighting the trends and contributions of a particular discipline, researcher or group of researchers, institution or country in relation to global scientific and technological progress (MACIAS-CHAPULA, 1998; STREHL & SANTOS, 2002).

According to Spinack (1998), the importance of the scientometric technique can be seen by analysing the list of possible applications below:

- identify trends and the growth of knowledge in different areas;
- measure the coverage of secondary magazines;
- identify the users of a subject;
- identify authors and trends in different disciplines;
- measuring the usefulness of selective information dissemination services;
- predict publication trends;
- identify the core magazines of a discipline;
- formulating budget procurement policies;
- adapting policies for discarding publications;
- study the dispersion and obsolescence of scientific literature;
- designing indexing processes, classifying and preparing automatic summaries;

- predict the productivity of publishers, individual authors, organisations, countries, etc.

To this end, scientometric approaches, through which science can be portrayed by the results achieved, are based on the notion that the essence of scientific research is the production of knowledge and that scientific literature is a component of this knowledge (MACIAS-CHAPULA, 1998).

CHAPTER 3

OBJECTIVES

3.1 GENERAL

- To characterise the scientific production of prostate adenocarcinoma, using the FISH technique as a molecular tool for detecting PTEN gene deletion, in the SCOPUS database, from 2005 to 2015.

3.2 SPECIFIC

* Verify the number of papers published in the stipulated research period;
* Quantify the types of publications (experimental articles, reviews, papers);
* Cite the main authors who have published the most and the areas involved in the topic studied;
* Check the journals that have published the most on the subject and their impact factor (IF);
* Identify the main institutional affiliations that have published the most on the subject;
* To see which countries have published the most articles on the subject;
* Note the main keywords used in the research carried out.

CHAPTER 4

MATERIALS AND METHODS

For the quantitative analysis of studies on PCa using the FISH technique as a cytomolecular tool in order to evaluate this methodology as a resource for identifying the deletion of the PTEN gene in this pathology, from 2005 to 2015, the scientometric method was used, which is based on the metric analysis of scientific production in a given area.

The studies were analysed using the keywords: *"poosaatic neoplasms* AND *pten* AND fsh". The use of the "AND" search operator retrieves only records containing both terms. The words used are MESH (*Medical Subject Headings)* descriptors, used by the Medline database on the PubMed portal, which uses standardised terminology to help define subjects and retrieve articles of interest.

In this study, the survey was carried out using the database published on the SCOPUS website, one of the largest internationally recognised databases for publications in scientific journals. It is maintained by the Elsevier Company, part of Reed Elsevier Group PLC, and has around 14,000 journals and 167 million scientific publications on various topics. SCOPUS began trading on 3 November 2004 and is now one of the most widely used and reliable scientific databases in the world (ELSEVIER, 2015). This database was used because of its comprehensiveness in terms of the number of publications and the quality of the journals indexed.

From the selected publications, the articles were identified and their *abstracts* read, taking into account the following information: (i) the year the article was published; (ii) the journal in which the article was published; (iii) the type of document published (experimental, review); (iv) the names of the authors of the work; (v) the area of knowledge in which it falls; (vi) keywords; (vii) the institutions to which the authors are affiliated; (viii) the countries where the studies were carried out and (ix) the Impact Factor (IF) of the journals that published the articles.

The *Journal Citation Reports* (JCR) is a database recognised for evaluating journals indexed in the Web of Science. It presents quantitative data that supports a systematic and objective review of journals covering various areas of knowledge, including the Biological Sciences. The JCR offers a perspective for evaluating and comparing journals through the accumulation and tabulation of citation counts and articles from practically every speciality in the field of science. It is an important tool for both researchers, who can determine where to publish their work and which journals to use in their research, and for librarians who analyse journal collections for acquisition (JCR, 2009).

The IF of the journals used in the analyses was obtained from the JCR for 2014. It provides a systematic and objective way of evaluating the world's leading research journals. It offers a unique perspective for evaluating and comparing journals by cumulating and tabulating citation counts and articles from virtually all specialities in the fields of science, social sciences and technology (JCR, 2009).

The IF identifies the average frequency with which an article in a journal is cited in a given year. It is calculated by dividing the number of citations in the year by the total number of articles published in the previous two years. This figure can be used to evaluate or compare a journal's relative importance with others in the same field or to visualise how often articles are cited to determine which journals are best for your research (REUTERS, 2015).

The results of the collection of articles included were tabulated and organised in an Excel® spreadsheet according to each research variable, as already mentioned. Tables were then drawn up separately with the study's descriptive statistics.

CHAPTER 5

RESULTS AND DISCUSSIONS

According to the survey carried out on the SCOPUS database, 29 papers were found from 2005 to 2015, distributed over these years, using the keywords *"poosaatic neoplasms* AND *pten* AND *fish"*. The low number of papers found using these terms is due to the specificity of the search keywords. However, four articles that did not contain the information sought were excluded. When analysing the number of articles published per year, using the terms cited, there is no uniform distribution, as shown in Figure 9, so that 2005 had no publications; 2007, 2008 2010 and 2015 had only one article per year of publication; in 2006 and 2009 two articles were published. The highest number of publications occurred in 2013 and 2014, when five articles were published each year. Three articles were published in 2011 and four in 2012. The years with the most publications were 2013 and 2014, with five publications each year. Over this specific period, there have been variations in the number of publications, the increase of which is related to the interest of researchers in this area of study and the fact that the number of publications is used as a measure to quantify the progress and evolution of science (VERBEEK *et al.*, 2002).

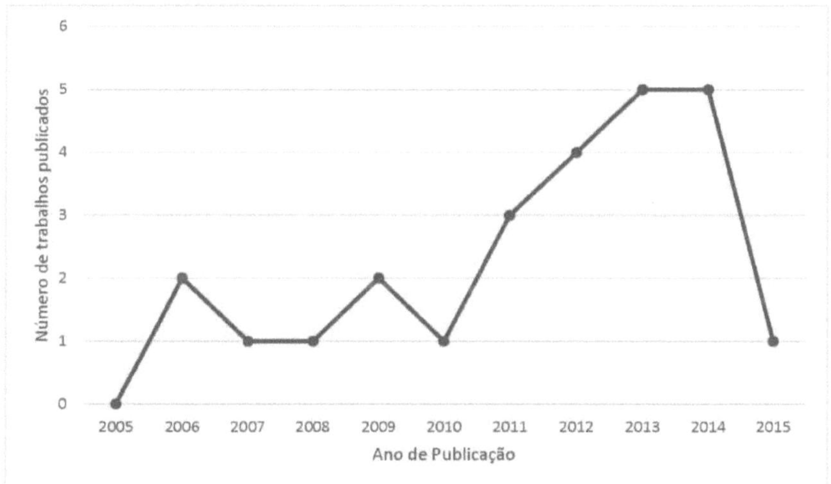

Figure 9. Distribution of the number of articles published in the field of genetics between 2005 and 2015, indexed in SCOPUS.

In Figure 10, the works analysed were mainly published as documents in the form of 24 articles (96%) and 1 review (4%) in 15 different journals. Thus, 96% of the published articles are experimental research papers, which shows that the studies carried out on this subject focus on practical and laboratory procedures rather than bibliographical reviews. And these experimental studies were guided by goals and strategies that sought new knowledge and answers about the fundamentals of phenomena and observable facts. Review studies had the lowest publication rate, with only one article (4 per cent). This tendency can be explained by the fact that they are time-consuming and scientists themselves believe that they do not bring attractive power. In fact, numerous studies in different areas report that theoretical studies are less frequent than experimental or descriptive studies (LIMA-RIBEIRO et al., 2007; CARNEIRO et al., 2008).

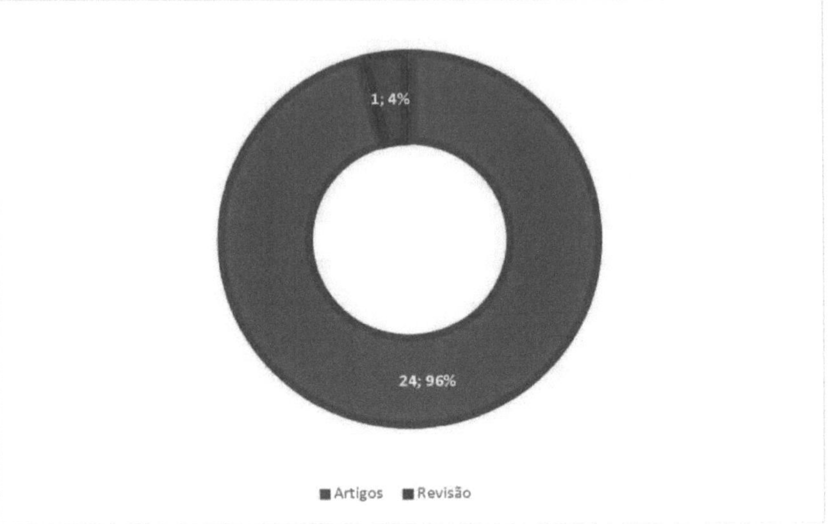

Figura 10.   Percentage and number of articles published according to the type of research.

When researching the subject, the articles were distributed across 15 journals. The number of articles published in each journal was then assessed. Three journals had a greater number of publications, the British Journal of Cancer, the Journal of Pathology and Modern Pathology published 3 articles each, which corresponds to 36 per cent of the total publications. Those that published two articles each were BMC Cancer, Cancer Epidemiology Biomarker, Plos One and Prostate, totalling 32%. The other eight journals published only one article each, totalling 32%. Figure 11 shows the journals and the number of articles published between 2005 and 2015.

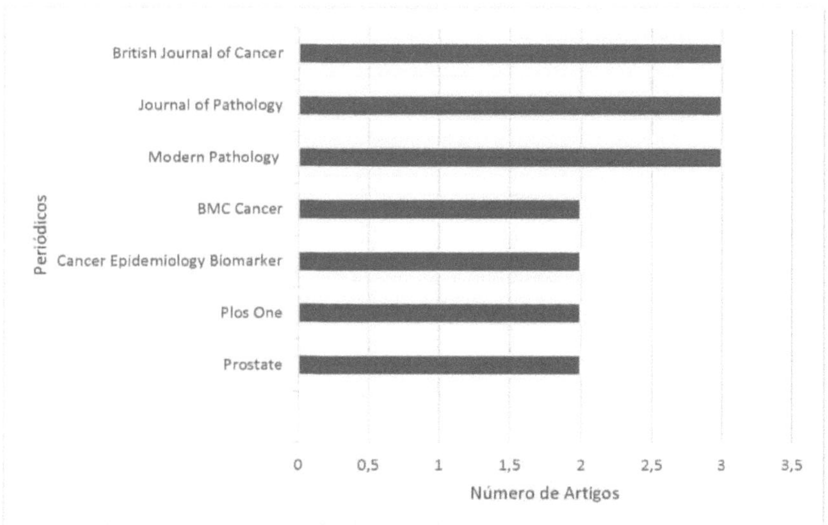

Figure 11. Number of articles published by journals analysed in the SCOPUS database.

The British Journal of Cancer provides a forum for doctors and scientists to communicate original research results that have relevance to understanding the aetiology of cancer and improving treatment and patient survival. Once accepted, articles are published online and in print, and their IF is 4.836 according to the 2014 JCR. The Journal of Pathology publishes important documents in pathology and experimental medicine in the form of the following documents: research articles, reviews, commentaries and perspectives, its IF is 7.429 according to the 2014 JCR. Modern of Pathology focuses mainly on aspects of modern pathology, publishing original articles and reporting on innovative clinical research, review articles, methods in pathology, special articles and case reports, its IF is 6.187 according to the 2014 JCR.

The authors were counted in full, with a total of 151 authors, attributing all their productivity to them, regardless of whether they were the main authors or collaborators. Of the authors who published articles on the subject studied over ten years, 8 published more than three articles, a frequency of 5%. Squire, J.A. and Yoshimoto, M. published seven articles each, totalling 1% of all published authors. Figure 12 lists the main authors who have published articles on the subject in the last ten years (2005-2015), showing that these two authors stand out in terms of the number of publications. In relation to the authors who have published three articles each, totalling 4%, there are six authors: Brewer, D.; Clark, J.; Cooper, C.S.; Joshua, A.M.; Soares, F.A. and Zielenska, M. In this same period, Bismar, T.A. and Chinnayan each published two articles.

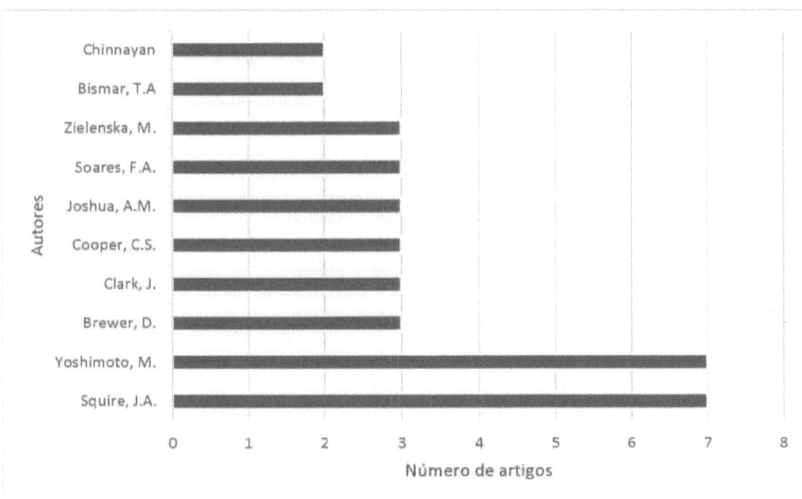

Figure 12. Names of the 10 authors who have published the most articles on the subject.

A total of 14 countries have researchers who have published on the subject, as shown in Figure 13, highlighting that this topic is of global interest due to its importance, as it is the type of cancer that most affects men and that its incidence is increasing due to the ageing of the population, associated with the improvement and greater popularisation of diagnostic systems. Adding up the number of articles published and the percentage of the four countries that have dedicated the most time to this issue are: the USA with 11 articles published (30%), Canada with 9 articles (24%) and Brazil and the United Kingdom with 3 articles each (8% each), totalling 70% of all production in relation to the other countries. This is due to the fact that in the United States prostate cancer represents a major public health challenge for its men, classified as the most commonly diagnosed cancer and one of the leading causes of cancer death in North American men (CHOUCAIR et al., 2012; BHALLA et al., 2013). This explanation also applies to other countries such as Canada, Brazil and the United Kingdom, which have seen an increase in PCa rates in recent years.

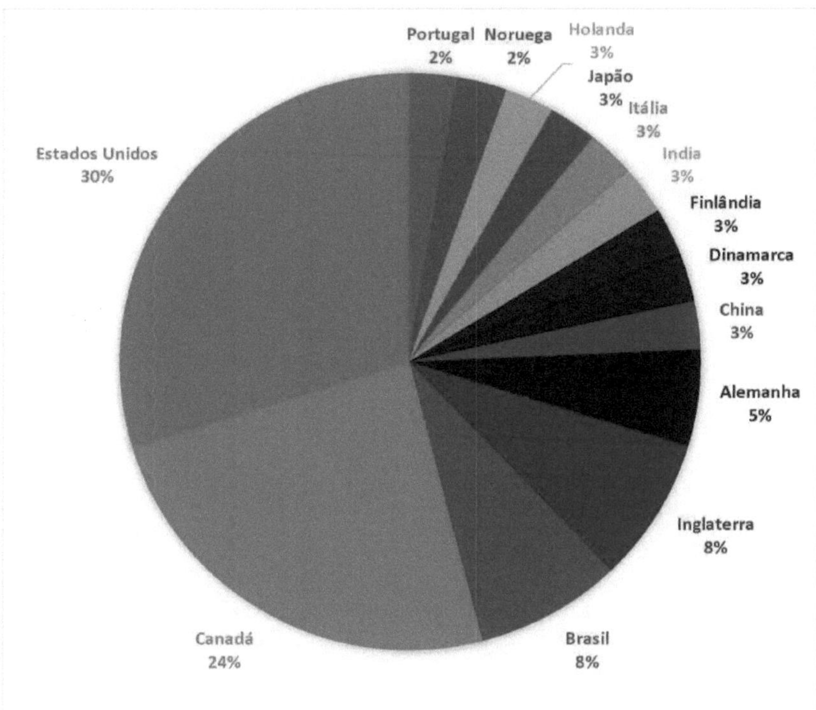

Figure 13. Description of the 14 countries that published during the research period (2005-2015).

The articles were published in four different scientific areas according to the SCOPUS classification, as shown in Figure 14. The area that published the most was Medicine, with 23 articles (56%), followed by Biochemistry, Genetics and Molecular Biology with 15 articles (36%), Agriculture and Biological Sciences with 2 articles (5%) and Pharmacology, Toxicology and Pharmaceuticals with only one article published (3%). This distribution of articles by area of study may be related to the fact that an article can be included in different areas of the study, since the naming of areas is not restricted to a single area. It is clear that the approach to the subject of prostate cancer related to the loss of the PTEN gene analysed by the FISH technique is a multidisciplinary interest, with its greatest impact in the field of Medicine. However, with advances in Genetics and Molecular Biology, the FISH technique has a large number of applications in this area. In this context, the advantage of this technique is that it allows maximum use to be made of tissue, which is difficult to obtain, and several different hybridisations can be carried out on the same tissue (QIAN & LLOID, 2003).

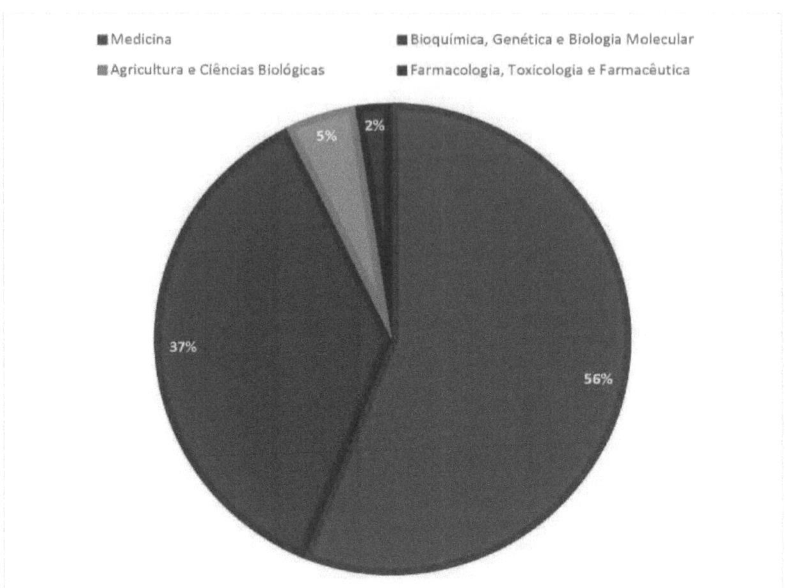

Figure 14. Main scientific areas of publication of studies related to PCa between 2005 and 2015, indexed in SCOPUS.

When checking the total number of keywords, which are simple terms or compound expressions used by the author to define the subject, in the articles included in the scientometric analysis, as shown in Figure 15, Prostate Cancer was the most used with 9 citations (14%), followed by the word PTEN with 8 citations (13%), the word FISH was cited 7 times (11%), Immunohistochemistry cited 5 times (8%) and Prognosis cited 4 times (6%) out of a total of 51 words related to the subject. These five words account for 52% of the total.

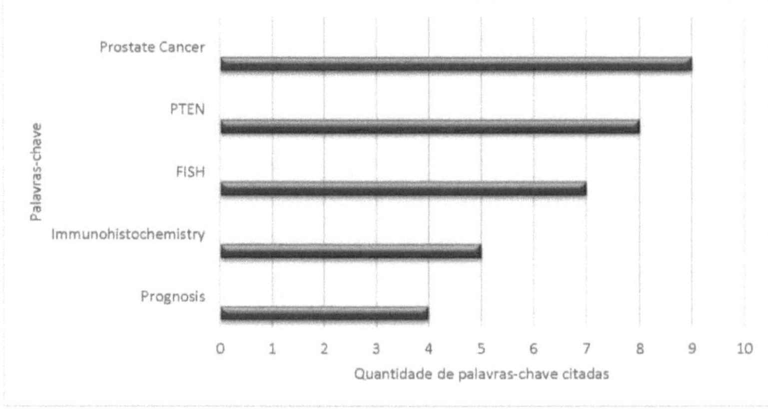

Figure 15. The five main keywords used in the publications relating to the research carried out.

A total of 93 different institutions were found by the authors' institutional affiliation, as shown in Figure 16. However, only 8 different institutions had their authors publish three or more articles, equivalent to only 25%

of the total number of publications. The institution that published the most, with 5 articles (4% of the total), was Princess Margaret Hospital University, which is considered the largest cancer study centre in Canada and one of the five largest in the world. Of the institutions that published 4 articles (9% of the total), *Queen's University, the University of Toronto and* the *University of Toronto Faculty of Medicine were* all Canadian, showing that this country, with a first-world economy, is interested in cancer studies, especially prostate cancer, and is investing satisfactorily in research. The *A.C. Camargo Hospital* stands out here in Brazil, having published 3 articles (3% of the total). This hospital is a non-profit private institution and is currently one of the largest integrated cancer centres in the world. Four different institutions published three articles, totalling 12%. However, the majority of institutions published one (65 per cent of institutions) or two articles (11 per cent of institutions).

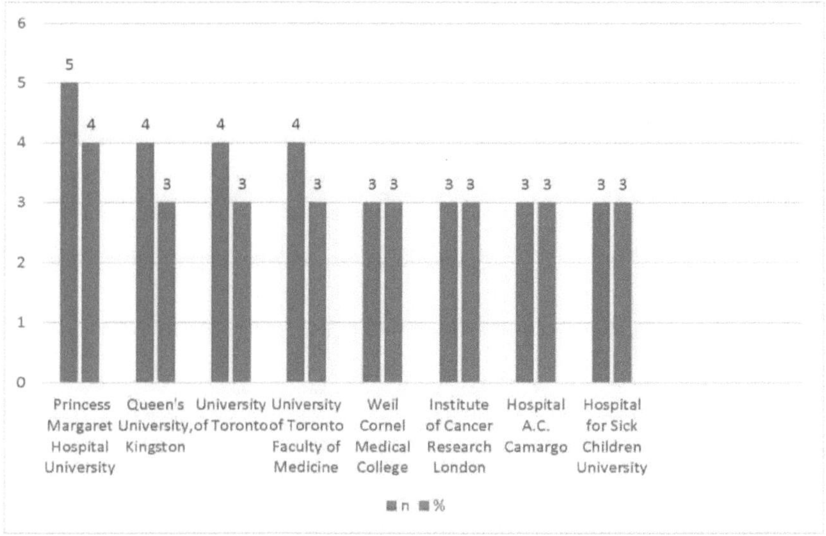

Figure 16. Main institutional affiliations of the authors who published the most on prostate cancer using FISH to analyse the PTEN gene.

This quantitative evaluation technique aims to measure the spread of scientific knowledge and the flow of information from a variety of perspectives. And this scientometric analysis of prostate cancer seeks to understand its importance as one of the cancers that most affects men today, relating it to the quantitative growth of its scientific production.

CHAPTER 6

FINAL CONSIDERATIONS

Following the scientometric analyses of the data collected, it was observed that the number of publications over the period studied, 2005 to 2015, varied over the ten years, with no uniformity in the publication of original articles on the proposed topic. However, in 2013 and 2014, the number of publications was higher than in other years. Most of these publications were in the form of original articles.

The journals that published the most on the topic of interest were the British Journal of Cancer, the Journal of Pathology and Modern Pathology, which accounted for 36 per cent of all publications. These journals are based in developed countries, confirming that these countries invest greater resources in cutting-edge research and are dominant in terms of their scientific prestige, with well-rated average impact factors.

It was found that the USA, Canada and Brazil were the countries that published the most articles on this subject, showing that not only developed countries are the main holders of scientific production in the world. Although Brazilian scientific production has grown significantly in recent years, the intellectual, social and economic impact of publications produced in Brazil remains low and still has a lot to grow. This is something that needs to change, with greater financial investment and qualified personnel, both from the government and private institutions, thus improving the credibility of Brazilian research institutions and researchers.

It was found that the main institutional affiliations of the authors who have published the most on prostate cancer using FISH to analyse the PTEN gene are Canadian, showing that this country invests well in research. However, one of these institutions is Brazilian, the A.C. Camargo Hospital, which is a private non-profit institution and is currently one of the largest integrated cancer centres in the world.

Therefore, it can be concluded that this scientometric analysis presented data from quantitative studies on prostate adenocarcinoma, highlighting the importance of each of the proposed objectives, seeking alternatives to improve the growth of science and the visibility of productions on this topic in the context of global scientific activity.

CHAPTER 7

BIBLIOGRAPHICAL REFERENCES

ALMEIDA, V.L.; LEITÃO, A.; REINA, L.C.B.; MONTANARI, C.A.; DONNICI, C.L. Cancer and specific cyclo-cellular and non-specific cyclo-cellular antineoplastic agents that interact with DNA: an introduction. **Química Nova.** 2005; Vol. 28. N.1: 148129.

ALLOTT, E.H.; MASKO, E.M.; FREEDLAND, S.J. Obesity and prostate cancer: weighing the evidence. **Eur Urol.** 2013;63(5):800-9.

ALTEKRUSE, S.F.; KOSARY, C.L; KRAPCHO, M. **Cancer Statistics Review.** 2010; 1975-2007. Bethesda, Md: National Cancer Institute.

AMERICAN CANCER SOCIETY: Cancer Facts and Figures 2015. Atlanta, Ga: American Cancer Society, 2015.

ANDRÉN, O.; FALL, K.; FRANZÉN, L.; ANDERSON, S.O; JOHANSSO, J.E.; RUBIN, M.A. How well does the Gleason score predict prostate cancer death? A 20-year follow-up in a population based cohort in Sweden. **J Urol.** 2006;175(4):1337-40.

AUS, G.; ABBOU, C.C.; PACIK, D.; SCHIMID, H.P. Guidelines on Prostate Cancer. **Eur Urol.** 2001; 40:97-101.

AVERBECK, M.A; BLAYA, R.; SEBEN, R.R. Diagnosis and treatment of benign prostatic hyperplasia. **Revista da AMRIGS.** 2010; Porto Alegre, 54 (4): 471-477.

AYALA, A.G.; RO, J.Y.; BABAIAN, R. The prostatic capsule: does it exist? Its importance in the staging and treatment of prostatic carcinoma. **The American Journal of Surgical Pathology.** 1989;13(1):21-27.

BERTRAM, J.; PEACOCK,J.W.; FAZLI, L.; MUI, A.L.; CHUNG, S.W.; COX, M.E.; MONIA, B.; GLEAVE, M.E.; ONG, C.J. Loss of PTEN is associated with progression to androgen independence. **Prostate.** 2006; 66:895-902.

BHALLA, R.; KUNJU, L.P.; TOMLINS, S.A. Novel Dual Colour Immunohistochemical methods for detecting ERG-PTEN and ERG-SPINK1 status in prostate carcinoma. **Modern pathology** : an official journal of the United States and Canadian Academy of Pathology, Inc. 2013;26(6):835-848.

BHAVSAR, A.; VERMA, S. Anatomic Imaging of the Prostate. **BioMed Research International.** 2014; 2014:728539.

BODEY, B.; BODEY, B. Jr.; KAISER, H.E. Immunocytochemical detection of prostate specific antigen expression in human primary and metastatic melanomas. **Anticancer Res.** 1997; 17:2343-2346.

BORGES DOS REIS, R.; CASSINI, M.F. Prostate Specific Antigen (PSA). In: ZERATI FILHO, M; NARDOZZA JÚNIOR, A.; REIS, R.B.(Orgs.). **Fundamental Urology.** São Paulo: Planmark, 2010. p. 189 - 194.

BOSTWICK, D.G. The pathology of early prostate cancer. **CA Cancer J Clin.** 1989; 39(6):376-393.

BOSTWICK, D.G.; GRIGNON, D.J.; HAMMOND, M.E.; AMIN, M.B.; COHEN, M.; CRAWFORD, D. Prognostic Factors in Prostate Cancer. College of American Pathologists Consensus Statement 1999. **Arch**

Pathol Lab Med. 2000; 124(7):995-1000.

BRAZIL. Ministry of Health. Health Care Secretariat. National Cancer Institute. TNM: classification of malignant tumours / translated by Ana Lúcia Amaral Eisenberg. 6. ed. - Rio de Janeiro: INCA, 2004.

BRAZIL. Ministry of Health. National Cancer Institute. Estimate / 2014 - Cancer incidence in Brazil. Rio de Janeiro: INCA; 2014.

BRAZIL. Ministry of Health/Secretariat of Health Care/Department of Regulation, Evaluation and Control/General Coordination of Information Systems. MANUAL DE BASES TÉCNICAS DA ONCOLOGIA - SIA/SUS - SISTEMA DE INFORMAÇÕES AMBULATORIAIS. 19ª Edition. Brasília. January 2015.

BRYANT, R.J.; HAMDY, F.C. Screening for prostate cancer: an update. **Eur Urol**. 2008; 53(1):37-44.

BUBENDORF, L.; SCHOPFER, A.; WAGNER, U. Metastatic patterns of prostate cancer: an autopsy study of 1589 patients. **Hum Pathol**. 2000; 31(5):578-583.

BUNKER, C.H.; PATRICK, A.L.; KONETY, B.R. High prevalence of screening- detected prostate cancer among Afro-Caribbeans: the Tobago Prostate Cancer Survey. **Cancer Epidemiol Biomarkers Prev**. 2002; 11 (8): 726-9.

CAMPOS, E. C. R.; FONSECA, F. P.; ZEQU, S. C. Analysis of PTEN gene by fluorescent in situ hybridisation in renal cell carcinoma. **Rev. Col. Bras. Cir.** 2013; 40(6): 471-475.

CARNEIRO, F. M., NABOUT, J.C., BINI, L.M. Trends in the scientific literature on phytoplankton. **Limnology**. 2008; 9: 153-158.

CARROL, P.; ALBERTSEN, P.C.; GREENE, K.; BABAIAN, R.J.; CARTER, H.B.; GANN, P.H. Prostate-specific antigen best practice statement: 2009 update. http://www.auanet.org/content/guidelines-and qualitycare/clinicalguidelines/main reports/psa09.pdf - Accessed on 18/Sep/2015.

CARVALHO, S. M. F. et al. Genetics of hereditary cancer. **Rev. Bras. de Canc.** 2009; v.55, n.3, p.263-269.

CARTER, B.S.; BEATY, T.H.; STEINBERG, G.D.; CHILDS, B.; WALSH, P.C. Mendelian inheritance of familial prostate cancer. **Proc Nati Acad Sci USA**. 1992; 89:3367-3371.

CHAN, J.M.; GIOVANNUCCI, E.L. Vegetables, fruits, associated micronutrients, and risk of prostate cancer. **Epidemiol Ver**. 2001; 23 (1): 82-6.

CHAN, J.M.; VAN BLARIGAN, E.L.; KENFIELD, S.A. What should we tell prostate cancer patients about (secondary) prevention? Current opinion in urology. 2014;24(3):318-323.

CHENG, I.; WITTE, J.S.; JACOBSEN, S.J. Prostatitis, Sexually Transmitted Diseases, and Prostate Cancer: The California Men's Health Study. Myer L, **ed. PLoS ONE**. 2010; 5(1):e8736.

CHOUCAIR, K.; EJDELMAN, J.; BRIMO, F.; APRIKIAN, A.; CHEVALIER, S.; LAPOINTE, J. PTEN genomic deletion predicts prostate cancer recurrence and is associated with low AR expression and transcriptional activity. **BMC Cancer**. 2012; 12:543.

CORRÊA, L.A.; BENDHACK, M.L.; SOYZA, A.A.O.; SABANEEFF, J. Prostate Cancer: Prognostic Factors. **Brazilian Society of Urology**. 2006; 30 supplement 01.

CRAWFORD, E.D.; ARAHAMSSON, P. PSA-based screening for prostate cancer: how does it compare with cancer screening tests? **Europ Urol.** 2008; 54(2):262-273.

DI FIORI, M.S.H. Atlas of Histology. 7ª Ed. Rio de Janeiro: Guanabara, 2000.

ELSEVIER. Scopus comes of age. http://www.elsevier.com/about/press- releases/science-and-technology/scopus-comes-of-age, 2014. Accessed on: 20 November 2015.

FELGUEIRAS.J.; SILVA,J.V.; FARDILHA, M. Prostate cancer: the need for biomarkers and new therapeutic targets. **Journal of Zhejiang University Science B.** 2014; 15(1):16- 42.

FERLAY, J.; SHIN, H.R.; FORMAN, D.; MATHERS, C.D.; PARKIN, D. Globocan 2008: cancer incidence and mortality worldwide. Lyon: International Agency for Research on Cancer; 2010.

FOSTER, C.S.; BOSTWICK, D.G.; BONKHOFF, H. Cellular and molecular pathology of prostate cancer precursors. **Scand J Urol Nephrol Suppl.** 2000; 34(1):19-43.

FREIRE, G.C.; PIOVESAN, A.C. Practical guide to urology / editors Donard Augusto Bendhack, Ronaldo Damião. 1.ed. Rio de Janeiro: SBU - Sociedade Brasileira de Urologia; São Paulo: **BG Cultural**, 1999. Chapter 15, pages 79-83.

GARÓFOLO, A.; AVESANI, C.M.; CAMARGO, K.G. Diet and cancer: An epidemiological view. **Rev. Nutr.** 2004; Campinas, 17(4):491-505, Oct./Dec.

GLEASON, D.F.; MELLINGER, G.T. Prediction of prognosis for prostatic adenocarcinoma by combined histological grading and clinical staging. **J Urol.** 1974; 111:58-64.

GOMES, R.; NASCIMENTO, E.F.; LEFS, R.; ARAÚJO, F.C. As arrancaduras da masculinidade: uma discussão sobre o toque rectal como medida de prevenção do câncer prostático. **Ciênc Saúde Colet.** 2008; 13(6):1975-84.

GRAY, I.C.; STEWART, L.M.D.; PHILLIPS, S.M.A.; HAMILTON, J.A.; GRAY, N.E.; WATSON, G.J.; SPURR, N.K.; SNARYL, D. Mutation and expression analysis of the putative prostate tumour-suppressor gene PTEN. **Br J Cancer.** 1998; 78 (10:1296-1300.

GREENE, K.L.; ABERTSEN, P.C.; BABAIAN, R.J.; CARTER, H.B.; GANN, P.H.; HAN, M.; KUBAN, D.A.; SARTOR, A.O.; STANFORD, J.L.; ZIETMAN, A.; CARROLL, P. Prostate Specific Antigen best practice statement: 2009 update. **J Urology.** 2009; 182(5):2232-2241.

GRIGNON, D.J. Unusual subtypes of prostate cancer. **Modern Pathol.** 2004; 17(3):316- 327.

GRONBERG, H.; DAMBER, L.; DAMBER, J.E. Familial prostate cancer in Sweden. A nationwide register cohort study. **Cancer.** 1996; 77:138-143.

GUTMAN, A.B.; GUTMAN, E.B. An "Acid" phosphatase occurring in the serum of patients with metastasising carcinoma of the prostate gland. **J. Clin. Invest.** 1938; 17:473-478.

HAAS, G.P.; DELONGCHAMPS, N.; BRAWLEY, O.W.; WANG, Y.C.; ROZA, G. The Worldwide Epidemiology of Prostate Cancer: Perspectives from Autopsy Studies. **Can J Urol.** 2008; 15(1): 3866-71.

HACKEL, C.; VARELLA-GARCIA, M. Interphase cytogenetics using fluorescence in situ hybridisation: an overview of its application to diffuse and solid tissue. **Braz. J. Genet.** 1997, Ribeirão Preto, v. 20, n. 1.

HAESE, A.; BECKER, C.; NOLDUS, J.; GRAEFEN, M.; HULAND, E.; HULAND, H. Human glandular kalikrein 2: a potential serum marker for predicting the organ confined versus non-organ confined growth of prostate cancer. **J Urol**. 2000; 163(5):1491-97.

HALLAL, C.I.A.; GOTLIEB, D.L.S.; LATORRE, M.R.D.O. Evolução da mortalidade por neoplasias malignas no Rio Grande do Sul, 1979-1995. **Rev Bras Epidemiol**. 2001; 4(3): 168-77.

HAN, B.; MEHRA, R.; LONIGRO, R.J. Fluorescence In situ Hybridisation Study Shows Association of PTEN Deletion with ERG Rearrangement during Prostate Cancer Progression. **Modern pathology**: an official journal of the United States and Canadian Academy of Pathology, Inc. 2009; 22(8):1083-1093.

HARA, M.; KOYANAGI, Y.; INOUE, T.; FUKUYAMA, T. Some physico-chemical characteristics of "seminoprotein", an antigenic component specific for human seminal plasma. Forensic immunological study of body fluids and secretion. VII. **NIhon Hogaku Zasshi**. 1971;25:322-324.

HAYTHORN, M.R.; ABLIN, R.J. Prostate-specific antigen testing across the spectrum of prostate cancer. **Biomark. Med**. 2011; 5:515-526.

HERNANDEZ, J.; THOMPSON, I.M. Prostate-specific antigen: A review of the validation of the most commonly used cancer biomarker. **Cancer**. 2004 ;101:894-904.

HESSELS, D.; SCHALKEN, J.A. Urinary biomarkers for prostate cancer: A review. **Asian J. Androl**. 2013.

HOROVITZ, D.D.G. Atenção aos defeitos congenitos no Brasil: propostas para estruturação e integração da abordagem no sistema de saúde [Doctoral Thesis] Rio de Janeiro: Instituto de Medicina Social, Universidade do Estado do Rio de Janeiro; 2003.

HOROVITZ D.D.G.; DE FARIA FERRAZ, V.E.; DAIN, S.; MARQUES-DE-FARIA, A.P. Genetic services and testing in Brazil. **Journal of Community Genetics**. 2013; 4(3):355-375.

HOVHANNISYAN, G.G. Fluorescence in situ hybridisation in combination with the comet assay and micronucleus test in genetic toxicology. **Molecular Cytogenetics**. 2010; 3:17.

HRICAK, H.; SCARDINO, P.T. Prostate Cancer. Contemporary Issues in Cancer Imaging. Cambridge, UK: Cambridge University Press; 2009.

HUNCHAREK, M.; HADDOCK, K.S.; REID, R.; KUPELNICK, B. Smoking as a risk factor for prostate cancer: a meta-analysis of 24 prospective cohort studies. **American journal of public health**. 2010; 100 (4):693-701.

IMPERATO McGINLEY, J.; GUERRERO, L.; GAUTIER, T. Steroid 5a reductase deficiency in man. An inherited form of male pseudohermaphroditism. **Birth Defects: Original Article Series**. 1975; 11(4):91-103.

NATIONAL CANCER INSTITUTE. Prostate Cancer Screening. November 2013

INCA - National Cancer Institute. What is cancer? Available at: http://www1.inca.gov.br/conteudo_view.asp?id=322. Accessed on: 27 August 2015.

PROSTATE AND URINARY INCONTINENCE INSTITUTE - IPIU - What it is and types of prostate cancer. Available at: http://institutodaprostata.com/cancro-da- prostata/o-que-e-o-cancro-da-prostata/. Accessed on: 31 August 2015.

JCR - Journal citation reports. Quick reference card: supported by Isi Web of Knowledge. [s.l]: Thomson

Reuters, 2009. Accessed on: 15 Oct 2015.

JENSEN, E. Technical Review: In Situ Hybridisation. **The Anatomical Record**. 2014; 297:1349-1353.

JEREZ-ROIG, J.; SOUZA, D.L.B.; MEDEIROS, P.F.M.; BARBOSA, I.R.; CURADO, M.P.; COSTA, I.C.C.; LIMA, K.C. Prostate cancer mortality projections in Brazil

Brazil: a population-based study. **Cad. Saúde Pública**. Rio de Janeiro. 2014; 30(11): 2451-2458.

JOHANSSON, J.E.; ANDRÉN, O; ANDERSON, S.O.; DICKMAN, P.W.; HOLMBERG, L.; MAGNUSON, A.; ADAMI, H.O. Natural history of early, localised prostate cancer. **JAMA**. 2004; 291(22):2713-19.

KOCH, M.O.; FOSTER, R.S., BELL, B.; BECK, S.; CHENG, L.; PAREKH, D. Characterisation and predictors of prostate specific antigen progression rates after radical retropubic prostatectomy. **J Urol**. 2000;164(3 Pt 1):749-53.

KOKSAL, I.T.; DIRICE, E.; YASAR, D.; SANLIOGLU, A.D.; CIFTCIOGLU, A.; GULKESEN, K.H.; OZES, N.O.; BAYKARA, M.; LUCELI, G.; SANLIOGLU, S. The assessment of PTEN tumour suppressor gene in combination with Gleason scoring and serum PSA to evaluate progression of prostate carcinoma. **Urol Oncol**. 2004; 22:307- 312.

KOLONEL, L.N. Fat, meat, and prostate cancer. **Epidemiol Ver**. 2001; 23 (1): 72-81.

LEE, C.H.; AKIN-OLUGBADE, O. KIRSCHENBAUM, A. Overview of prostate anatomy, histology, and pathology. **Endocrinology and Metabolism Clinics of North America**. 2011; 40(3):565-575.

LIMA-RIBEIRO, M.S., NABOUT, J.C., PINTO, M.P., MOURA, I.O., MELO, T.L., COSTA, S.S., RANGEL, T.F.L.V.B. Scientometric analysis in population ecology: importance and trends in the last 60 years. **Acta Scientiarum. Biological Sciences**. 2007; 29: 39-47.

LIU, W.; CHANG, B; SAUVAGEOT, J. Comprehensive assessment of DNA copy number alterations in human prostate cancers using Affymetrix 100K SNP mapping array. **Genes Chromosomes Cancer**. 2006; 45:1018-1032.

LOTAN, T.L.; GUREL, B.; SUTCLIFFE, S. PTEN Protein Loss by Immunostaining: Analytic Validation and Prognostic Indicator for a High Risk Surgical Cohort of Prostate Cancer Patients. **Clinical Cancer Research**. 2011; 17(20):6563-6573.

MACHADO, S. P.; SAMPAIO, H.A.C.; LIMA, J.W. Anthropometric characterisation of prostate cancer patients in Ceará, Brazil. **Rev. Nutr**. 2009. Campinas, v. 22, n.3, p. 367-376. Accessed on: 27 Aug. 2015. http://dx.doi.org/10.1590/S1415- 52732009000300007.

MACIAS-CHAPULA, C.A. The role of informetrics and scientometrics and their national and international perspective. **Information Science**. 1998. Brasília, v. 27, n. 2, p.134-140.

McCALL, P.; WITTON, C.J.; GRIMSLEY, S.; NIELSEN, K.V.; EDWARDS, J. Is PTEN loss associated with clinical outcome measures in human prostate cancer? **British Journal of Cancer**. 2008; 99(8):1296-1301.

MACLEOD, K. Tumour suppressor genes. **Curr Opin Genetc**. Dec. 2000; 10:81-93.

McMENAMIN, M.E.; SOUNG, P.; PERERA, S.; KAPLAN, I.; LODA, M.; SELLERS, W.R. Loss of PTEN expression in paraffin-embedded primary prostate cancer correlates with high Gleason score and advanced stage. **Cancer Res**. 1999; 59:4291-4296.

MARTIN, N.E.; MUCCI, L.A.; LODA, M.; DePINHO, R.A. Prognostic Determinants in Prostate Cancer. **Cancer journal (Sudbury, Mass)**. 2011; 17(6):429-437.

MEDICINE, U.S.N.L.O. Pten: Genetics, Home and Reference. 2007. Adapted and translated in October 2015 from: http://ghr.nlm.nih.gov/gene/PTEN.

MIGOWSKI, A.; AZEVEDO E SILVA, G. Survival and prognostic factors of patients with clinically localised prostate cancer. **Rev Saúde Pública**. 2010; 44(2):344-52.

MOREIRA, W. Literature Review and Scientific Development: concepts and strategies for preparation. Janus, São Paulo, year 1, no. 1, 2004. Available at: <http://www.nesc.ufg.br/uploads/19/original_Revis__o_de_Literatura_e_desenvolvime nto_cient__fico.pdf>. Accessed on: 20 Oct 2015.

MUGA, S.; SILVIA, H.; LAIA, A. Molecular alterations of EGFR and PTEN in prostate cancer: association with high-grade and advanced-stage carcinomas. **Modern Pathology**. 2010; 23, 703-712.

NAGLER, H.M.; GERBER, E.W.; HOMEL, P.; WAGNER, J.R.; NORTON, J.; LEBOVITCH, S. Digital rectal examination is barrier to population-based prostate cancer screening. Urology. 2005; 65:1137-40.

NASSIF, A.E.; FILHO, R.T. Immunohistochemistry expression of tumour markers CD34 and P27 as a prognostic factor of clinically localised prostate adenocarcinoma after radical prostatectomy. **Rev. Col. Bras. Cir.** 2010; 37(5): 338-344.

NATH, J., et al. "A Review of Fluorescence In Situ Hybridisation (FISH): Current Status and Future Prospects". **Biotech Histochem**. 2000.

NETTER, F.; OVALLE, W.K.; NAHIRNEY, P.C. Nettefs Essential Histology. 2014. 2ª Ed. Rio de Janeiro: **Elsevier**.

PARSONS, R. Human cancer, PTEN and the PI-3 kinase pathway. **Semin Cell Dev Biol**. 2004; 15:171-176.

PARTIN, A.W.; PIANTADOSI, S.; SANDA, M.G.; EPSTEIN, J.I.; MARSHALL, F.F.; MOHLER, J.L. Selection of men at high risk for disease recurrence for experimental adjuvant therapy following radical prostatectomy. **Urology**. 1995; 45:831-8.

PETO, J. Cancer epidemiology in the last century and the next decade. **Nature**. 2001; 411(6835): 390-5.

PINTO, A.C.; MACÉA, J.R. Surgical Anatomy of the Urinary and Genital Tracts. In: ZERATI FILHO, M; NARDOZZA JÚNIOR, A.; REIS, R.B.(Orgs.). Fundamental Urology. São Paulo: **Planmark**, 2010. p. 17-27.

POMPEO, A.C.L. Prostate Cancer. In: BENDHACK, A.D.; DAMIÃO,R. Practical Guide to Urology. 1. ed. Rio de Janeiro: SBU - Sociedade Brasileira de Urologia; São Paulo: **BG Cultural**, 1999. p. 162-172.

QI, M.; YANG, X.; ZHANG, F. ERG Rearrangement Is Associated with Prostate Cancer- Related Death in Chinese Prostate Cancer Patients. Yu J, **ed. PLoS ONE**. 2014; 9(2):e84959.

QIAN, X.; LLOYD, R.V. Recent developments in signal amplification methods for in situ hybridisation. **Diagn. Mol. Pathol.** 2003. 12: 1-13.

QU, X.; RANDHAWA, G.; FRIEDMAN, C., et al. A Three-Marker FISH Panel Detects More Genetic Aberrations of AR, PTEN and TMPRSS2/ERG in Castration-Resistant or Metastatic Prostate Cancers than in Primary Prostate Tumours. Tang DG, **ed. PLoS ONE**. 2013; 8(9):e74671.

REID, A.H..M; ATTARD, G.; BREWER, D. Novel, gross chromosomal alterations involving PTEN cooperate with allelic loss in prostate cancer. **Mod Pathol**. 2012; **United States and Canadian Academy of Pathology, Inc**.

REUTERS, T. Immediacy index. Journal Citation Reports. 2012a. Available at: <http://adminapps.webofknowledge.com/JCR/help/h_immedindex.htm#immed_index>. Accessed on: 15 Oct 2015.

RISINGER, J.I.; HAYES, A.K.; BERCHUCK, A.; BARRETT, J.C. PTEN/MMAC1 mutations in endometrial cancers. **Cancer Res**. 1997; 57: 4736-8.

ROHDEN, E.L.; AVERBECK, M.A. Localised prostate cancer. **Revista da AMRIGS**, Porto Alegre, 54 (1): 92-99, Jan.-Mar. 2010.

RUBIN, M.A.; DUNN, R.; KAMBHAM, N.; MISICK, C.P.; O'TOOLE, K.M. Should a Gleason score be assigned to a minute focus of carcinoma on prostate biopsy? **Am J Surg Pathol**. 2000; 24:1634-40.

SANTOS, R.N.M. Scientific production: Why measure? What to measure? **Digital Journal of Library and Information Science**, Campinas, 2003, v . 1, n. 1, p. 22-38.

SARDANA, G.; JUNG, K.; STEPHAN, C.; DIAMANDIS, E.P. Proteomic analysis of conditioned media from the PC3, LNCaP, and 22Rv1 prostate cancer cell lines: Discovery and validation of candidate prostate cancer biomarkers. J. **Proteome Res**. 2008; 7:3329- 3338.

SCHMITZ, M.; GRIGNARD, G.; MARGUE, C.; DIPPEL, W.; CAPESIUS, C.; MOSSONG, J.; NATHAN, M.; GIACCHI, S.; SCHEIN, R.; KIEFFER, N. Complete loss of PTEN expression as a possible early prognostic marker for prostate cancer metastasis. **Int J Cancer**. 2007; 120(6:1284-1292).

SCHODER, F.H.; HUGOSSON, J.; ROOBOL, M.J.; TAMMELA, T.L.J.; CIATTO, S.; NELEN, V.V.; KWIATKOWSKI, M.; LUJAN, M.H.; ZAPPA, M. Screening and

prostate-cancer mortality in a randomised European study. **N Engl J Med**. 2009; 360(13):1320-1328.

SCHULMAN, C.C.; ZLOTTA, A.R.; DENNIS, L.; SCHRODERE, F.H.; SAKR, W.A. Prevention of prostate cancer. **Scand J Urol Nephrol**. 2000; 205 (Suppl):50-61.

SMITH, R.A.; METTLIN, C.J.; EYRE, H. Cancer screening and early detection. In: Kufe DW, Pollock RE, Weichselbaum RR, editors. Holland-Frei Cancer Medicine. Canada: **BC Decker**. 2003.

SOUZA, L.M.; SILVA, M.P.; PINHEIRO, I.S. Um toque na masculinidade: a prevenção do câncer de próstata em gaúchos tradicionalistas. **Rev Gaúcha Enferm**. 2011. Porto Alegre (RS);32(1):151-8.

SPINAK, E. Indicadores cienciometricos. Ciência da Informação, Brasília. 1998. v. 27, n. 2, p. 141-148.

SROUGI, M. et al. Prostate diseases. **Rev. Méd**. 2008, São Paulo, v.87, n.3, p.166- 177.

STREHL, L.; SANTOS, C.A. Quality indicators of scientific activity. **Cienc. Hoje**. 2002, Rio de Janeiro, v. 31, n. 186, p. 34-39.

TAGUE-SUTCKIFFE, J. An introduction to informetrics. Information Processing and Management. **Oxford**. 1992. v. 28, n. 1, p. 1-3.

THOMPSON, I.M.; ANKERST, D.P.; CHI, C.; SCOTT, L.M.; GOODMAN, P.J.; CROWLEY, J.J.; PARNES, H.L.; COLTMAN, C.A. Operating Characteristics of Prostate-Specific Antigen in Men With an

Initial PSA Level of 3.0 ng/ mL or Lower. **JAMA**. 2005; 294(1):66-70.

THOMPSON, I.M.; ANKERST, D.P. Prostate - specific antigen in the early detection of prostate cancer. **CMAJ**. 2007; 176:1853-8.

UZOH, C.C.; PERKS, C.M.; BAHL, A. et al. PTEN-mediated pathways and their association with treatment-resistant prostate cancer. **BJU Int**. 2009; 104:556-561.

VANTI, N.A.P. From bibliometrics to webometrics: a conceptual exploration of the mechanisms used to measure the recording of information and the dissemination of knowledge. **Cienc. Inf.** 2002, Brasília, v. 31, n. 2, p. 152-162.

VELONAS, V.M.; WOO, H.H.; DOS REMÉDIOS, C.G.; ASSINDER, S.J. Current Status of Biomarkers for ·Prostate Cancer. **International Journal of Molecular Sciences**. 2013; 14(6):11034-11060.

VERBEEK, A. et al. Measuring progress and evolution in science and technology - I: The multiple uses of bibliometric indicators. **Int. J. Manag. Rev.**, Oxford, v. 4, n. 2, p. 179211. 2002.

VISAPAA, H.; SELIGSON, D.; HUANG, Y.; RAO, J.Y.; BELLDEGRUN, A.; HORVATH, S.; PALOTIE, A. Ki 67, gelsolin and PTEN expression in sarcomatoid renal tumours. **Urol Research**. 2003; 30:387-9.

WILSON, J.D.; GRIFFEN, J.E.; LESHIN, M.; GEORGE, F.W. Role of gonadal hormones in development of the sexual phenotypes. **Human Genetics**. 1981; 58(1):78- 84.

WOLF, A.M.; WENDER, R.C.; ETZIONI, R.B.; THOMPSON, LM.; D'AMICO, A.V.; VOLK, R.J., et al. American Cancer Society guideline for the early detection of prostate cancer: update 2010. **CA Cancer J Clin**. 2010; 60:70-98.

WORLD CANCER RESEARCH FUND. Food, nutrition and prevention of cancer: A global perspective. Washington: **American Institute for Cancer Research**. 1997. p.35- 71, 508-40.

WORLD HEALTH ORGANISATION. The World Health Report 1998: Life in the 21st century a vision for all. Geneva: **WHO**. 1998. p.61-111.

WU, C.P.; GU, F.L. The prostate in eunuchs. **Prog Clin Biol Res**. 1991. 370: 249-55.

WUNSCH FILHO, V.; MONCAU, E.J. Cancer mortality in Brazil 1980-1995: Regional patterns and temporal trends. **Rev Assoc Med Bras**. 2002; 48(3):250-57.

WUNSCH FILHO, V.; ANTUNES, J.L.F.; BOING, A.F.; LORENZI, R.L. Perspectives on research into social determinants in cancer. **Physis**. 2008; 18(3):427- 50.

YOSHIMOTO, M.; CUTZ, J.C.; NUIN, P.A., et al. Interphase FISH analysis of PTEN in histologic sections shows genomic deletions in 68% of primary prostate cancer and 23% of high-grade prostatic intraepithelial neoplasias. **Cancer Genet Cytogenet**. 2006; 169:128-37.

YOSHIMOTO, M.; CUTZ, J.C.; NUIN, P.A.; JOSHUA, A.M.; BAYANI. J.; EVANS, A.J.; ZIELENSKA, M.; SQUIRE, J.A. Interphase FISH Analysis of PTEN in Histologic Sections Shows Genomic Deletions are Present in 68% of Primary Prostate Cancer and 23% of High-Grade Prostatic Intra-Epithelial Neoplasia. **Cancer Genetics and Cytogenetics**. 2006. 169:128-37.

YOSHIMOTO, M.; CUNHA, I.W.; COUDRY, R.A., et al. FISH analysis of 107 prostate cancers shows that PTEN genomic deletion is associated with poor clinical outcome. **British Journal of Cancer**. 2007; 97(5):678-

685.

YOSHIMOTO, M.; JOSHUA, A.M.; CUNHA, I.W., et. al. Absence of TMPRSS2:ERG fusions and PTEN losses in prostate cancer is associated with a favourable outcome. **Modern Pathology**. 2008. 21: 1451-1460;

ZEQUI, S.C.; CAMPOS, R.S.M. Surgical Anatomy of the Urinary and Genital Tracts. In: ZERATI FILHO, M; NARDOZZA JÚNIOR, A.; REIS, R.B.(Orgs.). Fundamental Urology. São Paulo: **Planmark**. 2010. p. 205-213.

ZU, K.; GIOVANNUCCI, E. Smoking and aggressive prostate cancer: a review of the epidemiological evidence. **Cancer Causes Control**. 2009; 20 (10):1799-810.